# Temple and Sacred Courtyards in Khandesh

## *Biodiversity & Ethnobotany*

## THE AUTHORS

**Dr. S.M. Khare** received M.Sc. (Botany) and Ph.D. (Ethnobotany) in 2009 and 2018 respectively. An enthusiastic teacher and research worker in plant science. Published 11 research papers in national and international journals. Participated in 16 seminars, symposia, conferences and workshops in Botany. Has keen interest in various academic and scientific activities.

**Dr. Mrs. Shubhangi Pawar** obtained M.Sc., M.Phil. (Botany) and Ph.D. in (Ethnobotany) 1983, 1991 and 2004 respectively. Having teaching experience of 31 years. Published 74 research papers in national and international journals and also co-authored "Ethnobotany of Jalgaon District (M.S.)". Headed Department of Botany and worked as Vice-Principal of her college. Participated in 58 national and international seminars, symposia and workshops in Botany.

**Dr. D.A. Patil** obtained M.Sc.(Angiosperm Taxonomy Specialization) and Ph.D. (Botany) degree in 1978 and 1983 respectively from the then Marathwada University, Aurangabad (M.S.) Having teaching experience of 31 years at graduate and post-graduate levels and research experience over 40 years in various disciplines of Botany. Guided 16 students for research degrees in Experimental Taxonomy, Biodiversity, Agrobiodiversity, Ethnobotany and Etymology. Published over 400 research papers and 25 books by national and international publishers, few of them translated in foreign languages like European, Russian and American. Recipient of prestigious Prof.V.Puri Medal (2014) awarded by Indian Botanical Society and also Dr.E.K.Janaki Ammal Award (2020) by International Society of Ethnobotanists, apart from other 10 awards. A fellow of various scientific societies and associations (FLS, FES, FIAAT, FIBS). Worked as Editor or a Member of Editorial Board of Botanical Journals. Chairman, Board of Studies in Botany (2000-2005, 2010-2015) and Dean, Faculty of Science (2000-2005) of the then North Maharashtra University, Jalgaon (M.S.). His expertise is sought in various academic and scientific bodies and societies. Superannuated in August 2015 as a Professor in Botany and Principal of his college and still active in research.

# Temple and Sacred Courtyards in Khandesh

## *Biodiversity & Ethnobotany*

*– Authors –*

**Dr. S.M. Khare**

**Dr. Mrs. Shubhangi Pawar**

**Dr. D.A. Patil**

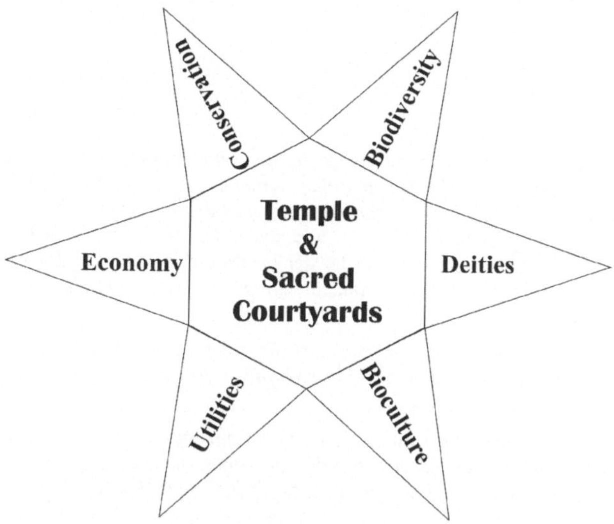


**2025**

**Scholars World**

*A Division of*

**Astral International Pvt. Ltd.**

**New Delhi – 110 002**

ISBN: 978-93-5919-994-8



*Published by* : **Scholars World**
*A Division of*
**Astral International Pvt. Ltd.**
– ISO 9001:2015 Certified Company –
4736/23, Ansari Road, Darya Ganj
New Delhi-110 002
Ph. 011-43549197, 23278134
e-mail: info@astralint.com
Website: www.astralint.com

*Respectfully Dedicated to*

**Dr. S.K. Jain**

***(Former Director, Botanical Survey of India)***

|||

**"Father of Indian Ethnobotany"**

**Who Devoted His Life for Biodiversity Studies and Geared up the Science of Ethnobotany in India**

# Preface

The lamp of 'Ethnobotany' was lit for the first time by Harshberger (1885). However, it was started rather a few decades late, in India. It is thought sometimes as a realm of new research or also part of the 'Economic Botany'. It has been defined variously in different times and its scope has now widened quietly touching both concrete and abstract relationship of mankind with the plantworld. The former relationships is, nevertheless, investigated extensively and intensively worldwide. Similar trend is also discernible in Indian subcontinent. It is needless to state that human sentiments are intricately interwoven with the deities in each societies, country or religion. However, floral elements associated with them have received rather cursory attention in ethnobotanical perspective. Moreover, biodiversity investigations in past have mostly overlooked in religious places or temple courtyards. At this backdrop, the present authors extended research to tap down ethnobotanical wisdom, bioculture *vis-à-vis* biodiversity studies in these hitherto neglected areas. Authors are delighted in presenting this sincere inventory in Khandesh region of the state of Maharashtra. It divulged ethnobotany and elements of biodiversity in religious or temple courtyards pertaining to various religions, tribal or non-tribal societies. As many as 263 plant species are recorded with due care, whether indigenous or alien ones. Besides the concrete use-reports, the sacred or ornamental plant species are noticed in view of economy and conservation *vis-à-vis* native bioculture in Khandesh region of Maharashtra (India). Indigenous applications of various botanicals deserve further research on scientific lines for the welfare of mankind.

***Dr. S.M. Khare***

***Dr. Mrs. Shubhangi Pawar***

***Dr. D.A. Patil***

# Contents

# Chapter 1

# Introduction

In real life, particularly in the developing world, there is a whole parallel knowledge based system, which is generated by people who live in the laboratory of life, based on their empirical wisdom and experience. There is an urgent need to preserve, protect and add value to it (Mashelkar, 2002). The talent of mankind to learn about the natural bioresources in ambience to his advantage has, *de facto*, made humans the most successful organism on blue planet. This talent and ability to innovate made him possible to build material culture and find out a distinct niche and home (Patil 2011). A man-plant relationship was clearly defined by Harshberger (1895) as 'the science denoting the uses of plants by the aboriginal people'. The term 'Ethnobotany' for this science was coined first by him. Since then, 'Ethnobotany' has been defined by various aspects and widened its scope (*cf.* Robbins *et al.*, 1916; Jones, 1941; Faulks, 1958; Schuites, 1963; Heiser, 1995) and few others. Harshberger *loc.cit.*) clarified purpose of ethnobotany as: (i) an aid to elucidating cultural positions of tribes according to the uses of plants, (ii) evidence of past distribution and trade routes and (iii) useful in suggesting 'the valuable properties they utilized in their wild life might fill some vacant niche in our own. Man-plant relationships fall under two broad categories *viz.* (i) concrete relationship which includes various kinds of uses of plants like for food, medicine, housing, basketry, tools *etc.* and (ii) abstract relationship includes social or cultural and religious or spiritual relationships with plants. *e.g.* weddings, festivals, dances, worships, rites, rituals, sacred groves, *etc.* Knowledge about all these is traditional and passed over past generations orally. Traditional knowledge survives generally among the indigenous or local communities. The science of ethnobotany has now emerged as multidisciplinary science and has linkages with botany, ecology, conservation, linguistics and pharmacology (Balick, 1996).

India has a very long socio-cultural history and heritage. The instinctive knowledge about medicinal uses of native plants was preserved by the Indian communities. This traditional knowledge was older than even the Vedas and Ayurveda. However, the 'Sanskritisation' or 'Refinement' pushed Indians away from such 'Unsanskritic practices' of following instincts (Manilal,2007). After European influx, scenario of Indian traditional or instinct knowledge changed, although it was taught by the Nature. Earliest record of documentation is hailed from Malabar region of India. Acosta (1578) in his own account included about 50 Indian medicinal plants. Later, Rheede (1678-1693) gave an excellent account of local medicinal plants from Malabar region of India. Roxburgh (1832) mentioned in his floristic surveys about local medicinal plants from South India. Watt's (1889-1896) 'Dictionary of the Economic Products of India' and 'The Commercial Products of India' (1908) are the later testimonies of the works on local uses of Indian plants. It was Bodding (1925, 1927, 1940) who published native plant medicines used by several tribes in India. After Indian independence, Dr. E.K.Janaki Ammal studied subsistence food plants of certain tribals of South India. She truly lit the lamp of 'Scientific Indian Ethnobotany' by starting an 'Ethnobotanical Section' at the Central Botanical Laboratory, Botanical Survey of India, Allahabad in 1960. Thus as an organised natural science, Ethnobotany in India is rather young, just some decades (Jain, 2001). Afterwards, there have been intensive and extensive explorations in this ream of science (Jain, 2015).

Ethnobotanical study is carried under two context, first, concreate relationship and second, abstract relationships. Most investigations have geared up the former one. As a result, the latter one is pushed on the descending spirals and is easily ignored. Shah (2008) clearly felt a need to investigate plants in abstract relationships. He certainly thus sought attention of ethnobotanical researchers towards cultural association of plants, in abstract form, which are in vogue are usually left out by most of the researchers (Shah, 2008.). Receently, Sood *et al.* (2005) put on record sacred and magico-religious plants of India. Sharma (2019) also paid attention on similar line in Assam (India). Mohanty *et al.* (1997, 2012) observed biodiversity in temple yards in the state of Odisha (Orissa), India. Similar work is also carried out by Dadhich and Dvivedi (2017) in the state Rajasthan (India), and Prabhakaran and Sabari Lakshmi (2017) in Tamil Nadu State (India).

## Previous Explorations in Maharashtra

Ethnobotanical research was initiated by Vartak and his co-workers in Maharashtra. They also emphasised sacred plants and groves. Pawar and Patil (2008) published ethnobotanical investigations in Jalgaon district (Maharashtra). However, the temple yards and sacred places have remained untouched to date. Khare *et al.* (2021a,b; 2022) published some accounts on holy places in Khandesh.

The literary resume clearly warranted to undertake inventory on plant species in sacred and temple courtyards in Khandesh region of Maharashtra in ethnobotanical perspective.

## Reasons for Undertaking Research

1. The accelerated loss of traditional or indigenous plantlore has to be focused before it disappears from the scene.
2. The rate of acculturation is faster than ever on account of modernisation and globalisation (This obviously results in erosion of traditional knowledge).
3. Biodiversity-rich nations (like India) are now considering indigenous plantlore as a part of their national heritage.
4. Natural product research is again on the rise. A number of new molecules or drugs of plant origin have awakened interest in using indigenous knowledge to guide drug discovery efforts.
5. Alternative approaches to healing have become more open in recent times. Green chemicals are being prioritised over synthetic ones.
6. Earlier distribution and structure of forests in Satpura ranges in Khandesh region was studied (Mahabale and Karnik, 1960). Also, observations on grasslands of Khandesh were made (Puri *et al.*, 1960). As far as the present area under study is concerned, classic floristic (biodiversity studies) have been fairly completed by Patil (2003) and Kshirsagar and Patil (2008). Pawar and Patil (2008) also paid attention to ethnobotany of Jalgaon district (exclusively), however, studies in abstract relationships appeared rather overlooked. To fill in this lacuna and at the suggestion of senior authors Dr.D.A.Patil and Mrs.Dr.S.B.Pawar, the present topic of research as an exclusive topic of research was extended over three districts of region which is comprised of Dhule, Nandurbar and Jalgaon districts in the northern part of Maharashtra (India).

# Chapter 2

# Geography and Geology

Khandesh region of Maharashtra is comprised of three districts *viz.*, Jalgaon, Dhule and Nandurbar. It is situated at the northern border of Maharashtra (Map-I). It is formerly called 'Khandesh'. Khandesh was divided in 1906 into two district *viz.*, West Khandesh and East Khandesh which are renamed after the district headquarters as Dhule and Jalgaon district respectively. Dhule district is again divided in two district Dhule and Nandurbar. Khandesh region lies between 22° 38′ to 22° 3′ Northern latitude and 73° 47′ to 76° 28′ East latitude and located in the upper Tapi basin and Deccan trap covering all most the whole area (Maps I-II). This region presently comprises 25 administrative tehsils *viz.* Jalgaon, Jamner, Erandol, Dharangaon, Bhusawal, Bodwad, Yawal, Raver, Muktainagar, Amalner, Chopda, Parola, Pachora, Chalisgaon, Bhadgaon (Jalgaon District); Dhule, Sakri, Shirpur, Shindkheda (Dhule District); Nandurbar, Taloda, Navapur, Akkalkuwa, Dhadgaon and Shahada (Nandurbar District).

## People

The population in Khandesh region according to census of 2012 is 6,702,346 spread over three districts and 25 tehsils. The tribal population is 1,739,089. Nandurbar district predominantly tribal with 72.29 per cent of the total population of the district. While in Dhule district, it is 29.97 per cent and Jalgaon 11.84 per cent of the total population respectively. The tribal villages or hamlets mostly belong to Nandurbar, Sakri, Navapur, Shahada, Taloda, Dhadgaon, Akkalkuwa, Chopda, Yawal, Raver, Erandol, Jamner, *etc.* Hindus form the largest proportion of this population. Muslims being the next largest community. Jains, Buddhists, Christians, Jews, Sikhs and Parsees also reside.

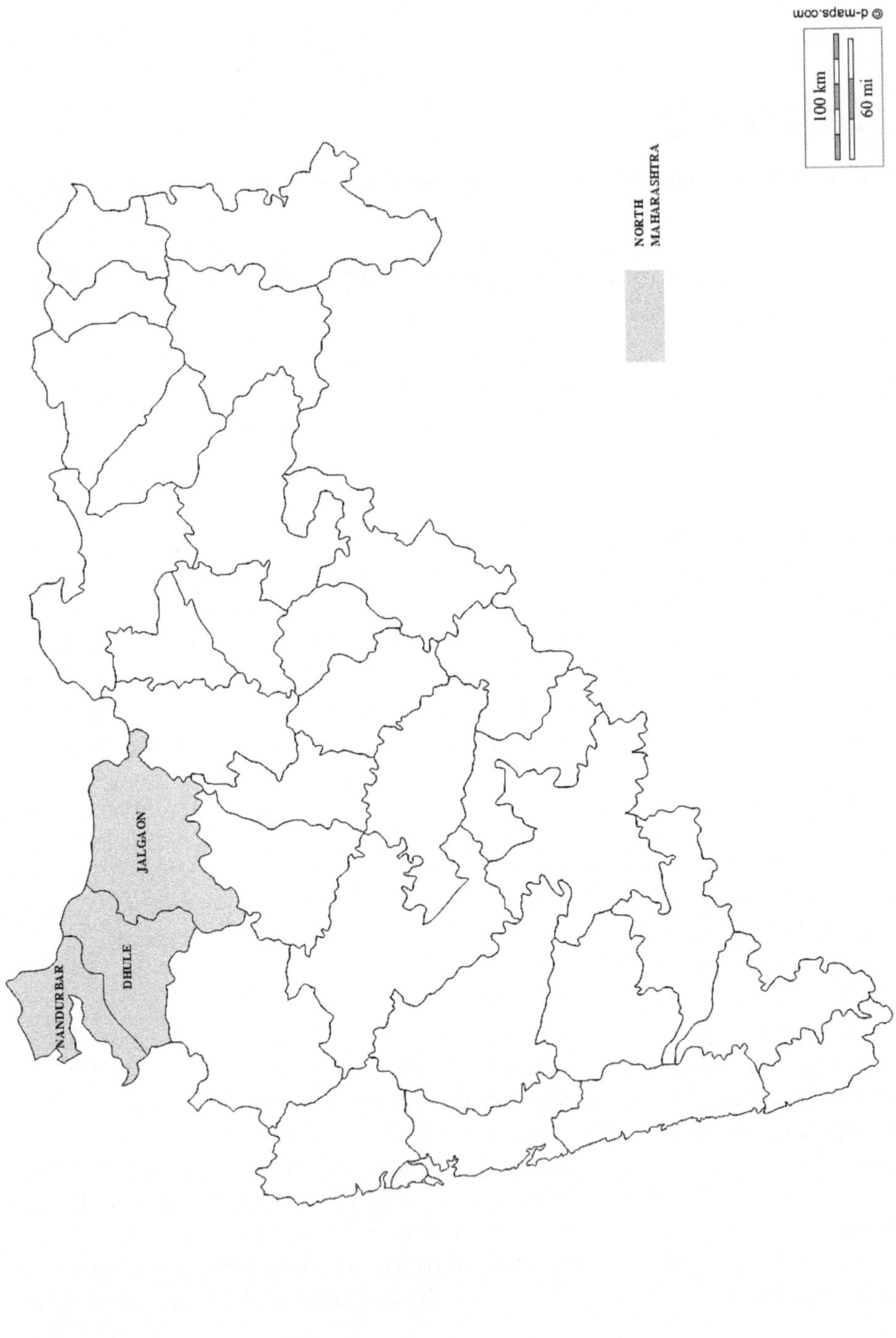


Map I: Khandesh Region in State of Maharashtra (India).

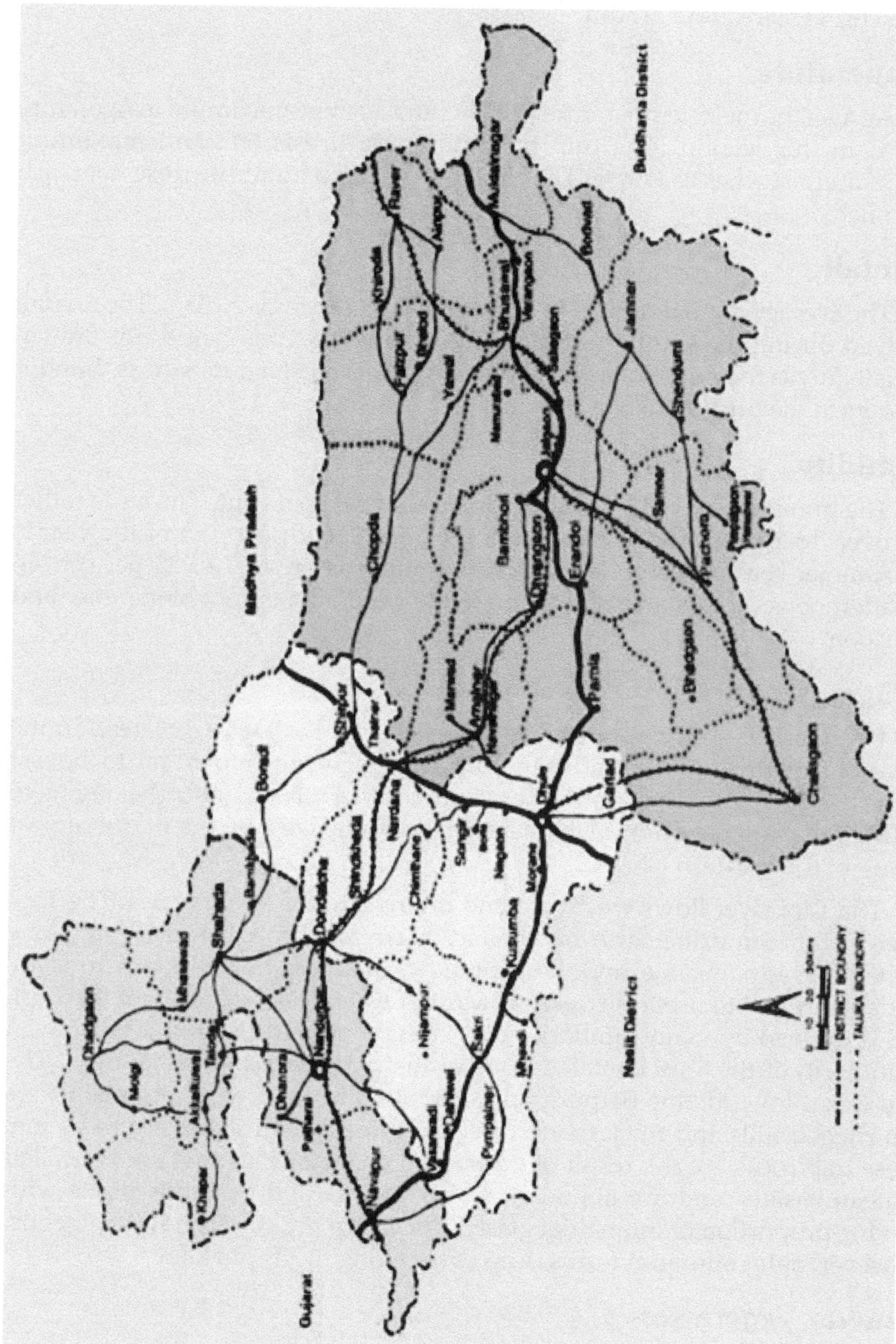


**Map II: Map of Khandesh Region.**

The aborigines belong to tribes *viz.*, Bhil, Pawaras, Kokanis, Tadvis, Kokana, Mavachi, Vanjara and Paradhi.

## Temperature

An average mean temperature is 11.9°C and average maximum temperature 29.8°C in this region. The minimum temperature is 6°C and maximum temperature reaches to 47.8°C. The climate is dry and humidity dose not reach too much.

## Rainfall

The average annual rainfall in Khandesh is 674.00 mm (26.53″). The rainfall received during the south-west monsoon is nearly 88 per cent of the annual rainfall. July is the rainiest month. Some rainfall is received mostly as thunder showers in the post-monsoon season.

## Humidity

The humidity except, the monsoon, is above 70 per cent. The air is rather dry over the region during the rest of the year. The driest part of the year is the summer season when the relative humidity is only 20 to 25 per cent in the afternoons. Winds are light to moderate generally, except of summer and monsoon.

## Geography and Soil

Khandesh lies on the Northwestern corner of the Deccan plateau in the valley of Tapi River. It is bound to the north by the Satpura mountain, to the east by the Berar (Vidarbha) region, to the south by the Hills of Ajanta (belonging to the Marathwada region of Maharashtra) and to the west by the northernmost ranges of the Western Ghats.

The Tapi river flows westward and drains into the Arabian Sea. The Tapi receives thirteen principal tributaries in its course through Khandesh. None of the rivers is navigable, and the Tapti flows in a deep bed which historically made it difficult to use for irrigation. Most of Khandesh lies south of the Tapi, and is drained by main tributaries *viz.*, Girna, Bori and Panjhra. The alluvial plain north of the Tapi contains some of the richest tracts in Khandesh. The land raises towards the Satpura hills. To the north and west, the plain rises into rugged hills and thick forests. Deccan traps extend in the entire region. These trap rocks are the result of outpoured enormous lava. These are called 'Plateau basalts' and are abundant in labrodorite and enstatite-augite with varying proportion of interstitial glass, which give rise to secondary minerals. However, older alluvium forms deep cover along the Tapi valley.

## General Vegetation

The forests in Khandesh belong to the tropical, dry deciduous type wherein *Tectona grandis, Hardwickia binata, Boswellia serrata, Madhuca indica* are the predominant species. The vegetation varies with the changes in altitude, aspect and rainfall. As these factors are variable, various subtypes of forest are met within the region. The different subtypes of the forest may be considered into I)*Tectona grandis* type, II) *Hardwickia binata* type, III)*Boswellia serrata* type, and IV) Scrub type.

1. ***Tectona grandis* type:** This type, in general, is found in well protected forest areas. The best type of this forest are found in Manudevi (Yawal), Eklagna (Erandol), Satrasen, Unapdev area (Chopda), Toranmal (Dhadgaon), Laling (Dhule), *etc.* The other trees generally observed are *viz., Anogeissus latifolia, Boswellia serrata, Garuga pinnata, Madhuca longifolia, etc.* During August to October, the common herbs are *viz., Cassia pumila,* species of *Indigofera, Alysicarpus, Tricodesma, Justicia etc.*
2. ***Hardwickia binata* type:** The chief localities of this type of forest are the eastern part at 'Pal' (Raver), western part of Lasur (Chopda). The other common trees in this type of forest are *viz., Hardwickia binata, Tectona grandis, Garuga pinnata, Mitragyna parvifolia, Serculia urens, etc.* The climbers and lianas found in this type are *Ampelocissus latifolia, Combretum ovalifolium, Hemidesmus indicus, etc.*
3. ***Boswellia serrata* type:** This type of forest is chiefly found in Padmalaya (Erandol) and Sakri region. This plant species is very common in this region. The other trees found in this region are *viz., Madhuca longifolia, Anogeissus latifolia, Acacia chundra, Acacia leucophloea, Cordia dicotoma, etc.* Common shrubs or small trees are in this forest are *viz., Mimosa hamata, Grewia flavescens, Euphorbia neriifolia, etc.*
4. **Scrub type:** In this type, the area is covered with open grasslands with few scattered shrubs and trees. These occur in the plain areas near Padmalaya, Ekalagna, Pal (Jalgaon District), Laling fort (Dhule District). A large number of rivulets and streams drain in these areas. The common trees and shrubs in this area are *viz., Dolichandrone falcata, Balanites aegyptiaca, Lantana camara, Acacia leucophoea, Capparis sepiaria etc.*
5. **Crops:** North Maharashtra region is predominantly agrarian and crops mainly belong to cereals, psudocereals, millets, pulses, oil crops, fibber crops, sugarcane, spices, edible fruits and vegetables. Cereals mainly include wheat, maize, rice and millets such as jwar (sorghum) bajara (pearl millets), foxtail millets, and pulses are also common *e.g.* blackgram, greengram, horsegram, pigeon pea, *etc.* Banana, sugarcane and cotton are cash crops. Tribals have small land holdings and have to depend also partly on forest resources. Domestic cattle are reared by farming communities.

# Chapter 3

## Methodology Adapted

The present investigation was extended by undertaking following steps:

I. Information was gathered through personal interviews of different people regarding sacred plants, beliefs, utilization of plants in worships, festivals, different conservation strategies, plant parts used, purpose of use, methods of processing and preparation of drug, its dosage and administration. Apart from the rural folks and urban elites, tribals such as Bhils, Pawaras, Vanjaras, Tadvis, Mavchi and Kokanies (Kokanas), *etc.* inhabit in this region. They belong to religions *viz.*, Hindu, Muslim, Baudha, Jain, Parshi, *etc.*

II. During this study the rural and tribal people were regularly interviewed and also the old people, Maharaj (Priest), Pujari (Worshiper), Brhamans, Vaidu, Buva, Bhagats, different Shepherds and heads of tribal communities. The interview was totally focused on the plants which have sacred importance in their society and few wild species in temple courtyards. Regular enquiries and observations during folk gatherings, pilgrims and ceremonies were made

III. The teachers, social workers, forest guards and rangers also helped to contact various informants of interest in the study region. Different sacred forests, ashrams, sacred sites, sacred gardens, temple groves, hamlets (Padas) were visited in different seasons to collect information. Most of the senior and knowledgeable informants also accompanied during the field study and plants were collected for voucher specimens. To confirm the authenticity of the data obtained during field work, repeated verification from different informants in different areas was completed. Supplementary information is collected to suit the background social information (Appendix:I-VIII).

IV. The reliable information was thus obtained in order to make this information useful to general public in future. Special attention was also paid to record local names of plants. Festivals, dances, ceremonies, social gatherings, pilgrims and temples were visited on some special occasions to note down the plant part used for different religious and concrete purpose. The traditional beliefs, taboos, restrictions and the folklore pertaining to the sacred plants were collected from different religions. This was also done enquiring famous Maharaj, Buwas, Pujaries or Bramhins associated with temple trusts in different cities and villages. Their actual wording, meaning and occasions on which they informed or used were enquired carefully.

V. During field studies plant specimens, their parts or products were gathered. Each collection, place of collection was noted in field diaries. The identification was done using district, state, regional and national floras. The family of the plants was deciphered by using Patil's key to the angiospermic families of North Maharashtra (1992). The floras such as: (a) The Flora of British India Vol. I-VII by Hooker, J.D. (1872-1897), (b) Flora of Maharashtra State: Monocotyledons by Sharma B.D. *et al.* (1996), (C) Flora of Maharashtra State: Dicotyledons. (2000, 2001) Vol. I and II by Singh *et al.* (d) Flora of Marathwada Vol I and II by Naik V.N. (1998) (e) Flora Dhule and Nandurbar Districts (2003) by Patil D.A. (2003), Flora of Jalgaon District by Kshirsagar, S.R. and D.A.Patil (2008) and few other floras published from the state of Maharashtra were of particular help in decipherment of the species. Exotic status is provided consulting Reddy (2008) and Patil (2021), Bailey (1949), apart from the floras given herewith and as special references mentioned against certain plant species in the text (Table-3).

VI. Photographs pertaining to plants or parts, tribal informants, forests, temples knowledgeable seniors, *etc.* have been provided to authenticate the information (Photographs 1-39). The relevant information regarding religion, caste, tribe, ceremonies, rites, utilities were mostly observed personally during the botanical outings. This study focused wild as well as cultivated plants used by the tribals, rural and urban and semi-rural areas of the region. The urban population was also of particular help on such occasions to have accurate information. Many house wives in urban societies and particularly the elder ladies or grandmothers were of immense help in case of domestic uses, religious rites and rituals. Regular field trips were undertaken in the tribal, rural and urban areas during 2013 to 2017. The information regarding ethnobotanical importance of medicinal plants was confirmed by frequent inquiries.

**Photograph 1: Holi Festival: Dried Fruits of Dudhi-bhopala (*Lagenaria siceraria*) decked during Dance.**

**Photograph 2: Patnadevi (Jalgaon district): use Naral (*Cocos nucifera*) in worship.**

**Photograph 3: Hanuman Khora, Bambhori (Jalgaon district): site covering in canopy of Neem (*Azardirachta indica*) and Behada (*Terminalia bellirica*)..**

**Photograph 4: Neem (*Azardirachta indica*) and Tribals from Ranipur (Nandurbar district).**

**Photograph 5: Dr. Khare interacting with villagers of Kalapani (Nandurbar district).**

**Photograph 6: Dr. Khare holding discussion with farmers of Shirpur region (Dhule district).**

**Photograph 7: Neem (*Azardirachta indica*), Wad (*Ficus benghalensis*) Valziri, Chalisgaon (Jalgaon district).**

**Photograph 8: Discussion with Vaidu (Medicine Man) Chandrasing Baba Leghapani (Nandurbar district).**

**Photograph 9: Religious place Astambha (Nandurbar district).**

**Photograph 10: Dr. Khare collecting plants at Manudevi Temple (Jalgaon district).**

**Photograph 11: Field visit at Langada Amba (Tribal Hamlet) (Jalgaon district).**

**Photograph 12: A lady worshiping a tree of Wad (*Ficus benghalensis*).**

Photograph 13: Mazar of Muslim Saint. Padmalaya roadside (Jalgaon district).

Photograph 14: Us (*Saccharum officinale*) plant used for making the taint of God Khandoba during Gondhal Jagran ceremony.

**Photograph 15: Decorative flowers along with Pimpal tree (*Ficus religiosa*) leaves used in the worship of Lord Buddha at Bhusawal (Jalgaon district).**

**Photograph 16: Tulsi (*Ocimum tenuiflorum*) *at* Sai mandir, Kasoda (Jalgaon district).**

**Photograph 17: "Jatra" at Patnadevi (Jalgaon district).**

**Photograph 18: Maharukh (*Ailanthus excelsa*) and Wad (*Ficus benghalensis*), Changdev, Muktainagar (Jalgaon district).**

**Photograph 19: Leaves of Bel [*Aegle marmelos* (L.) Corr], flowers of Zendu (*Tagetes erecta* L.) and fruits of Limb [*Citrus aurantifolia* (Christm. and Panz.)] used during worship.**

**Photograph 20: Neem (*Azardirachta indica*) Rail Maruti, Dharangaon (Jalgaon district).**

**Photograph 21: Rui (*Calotropis gigantea*), Shravan Bal Temple, Hartale, Muktainagar (Jalgaon district).**

**Photograph 22: Naral (*Cocos nucifera*), Nirgudi (*Vitex negundo*) *etc.* Shivdham, Parola (Jalgaon district).**

**Photograph 23: Use of turmeric powder (*Curcum longa*) at the temple of Khandoba, Nagardeola (Jalgaon district).**

**Photograph 24: Pimpal (*Ficus religiosa*), Pohi, (Jalgaon district).**

**Photograph 25: Pilligrim of Kapileshwar, Amalner (Jalgaon district).**

**Photograph 26: *Ficus religiosa* at Kapileshwar Temple, Mudawad, Tq.Amalner (Jalgaon district).**

**Photograph 27: View of Manudevi Temple (Jalgaon district).**

**Photograph 28: Buddha Vihar, Amalner (Jalgaon district).**

**Photograph 29: Pirbaba Dargah, Lone (Jalgaon district).**

The observations are based on ethnobotanical and floristic surveys conducted in urban, rural and tribal areas of Jalgaon, Dhule and Nandurbar districts. Mostly the temple yards in this region are visited. The plant species are enumerated alphabetically with their correct botanical names, family, vernacular names, sacred aspect and brief note on administration of medicine. Critical notes (C.N.) are provided separately below.

VII. Background information of Jalgaon, Dhule and Nandurbar districts was gathered w.r.t. geography, forest areas, people, rainfall, temperature, crops *etc.* through consultation of published gazetteers.

a) Local economy was also studied during these visits and discussed with the local informants. Quantification and local prices were also enquired in some cases of interest. In case of sacred plants, the name of deities, the people who worship and occasion of festivals or name of worships have been noticed carefully.

b) In view of conservation, the plant species conserved by various communities was observed. Practices related to conservation were enquired. Benefits accrued due to these conservation practices was discussed pertinently during field studies.

# Chapter 4

# Systematic Enumeration

## 1. *Abelmoschus esculentus* (L.) Moench.

**Family:** Malvaceae

**Local Name:** Bhendi

Stout, annual; large leaves nearly 8-10 cm or sometimes more; flowers axillary, solitary; capsules beaked, hairy, becoming glabrous, woody at maturity, seeds pale brown, minutely striate.

**Sacred Aspect**

I. It is used an offering to goddess Mahalakshmi.

**Medicinal Uses**

I. A cup of decoction of root is topically used as antiseptic on wounds and cuts.

II. A spoonful of fruit powder is used as abortifacient for three consecutive days.

**Miscellaneous Uses**

I. Fruits are used as vegetable.

**C.N.:** It is exotic, native of Africa and cultivated for edible fruits.

## 2. *Abelmoschus manihot* (L.) Medik.

**Family:** Malvaceae

**Local Name:** Kasturi bhendi, Ran bhendi

Herb with slender tap root; leaves are alternate, rough, hairy and heart-shaped, 3 to 5 lobes and grows up to 15cm long; flowers axillary, solitary, yellowish or pinkish; fruits up to 6 cm. long.

**Sacred Aspect**

I. Bhil tribe of Dhadgaon and Toranmal region believed that seeds if stored in house are auspicious.

II. Occasionally the powder of seeds is used for the preparation of food for offering to god.

**Medicinal Uses**

I. Two teaspoon seed powder is mixed with a cup water and a teaspoon honey. It is given to a patient twice to thrice a day in the problem of indigestion until it is cured.

II. Bark paste is topically used as an antiseptic on cuts and wounds.

III. A teaspoon powder of bark is mixed with coconut oil. The paste is used topically on rheumatism and swellings.

**Miscellaneous Uses**

I. Fiber is used for domestic purpose.

## 3. *Abutilon indicum* L.

**Family:** Malvaceae

**Local Name:** Sikka, Atti

Much branched, annuals or perennials, shrubby; leaves ovate or suborbicular, subtrilobate, glaucous beneath; flowers orange yellow, solitary, axillary; seeds reniform-ovoid, blackish-brown.

**Sacred Aspect**

I. Fruit of the plant is used in the worship of gramdevta in Pawra tribe.

II. Presence of the plant in the forest is consived as sacred thing by tribals.

**Medicinal Uses**

I. Oil extracted from the roots is used topically on joint pain.

II. A teaspoon powder of stem is mixed with a cup boiled water with two teaspoon sugar; this is prescribed twice a day in the treatment of diabetics.

III. A teaspoon leaf powder is mixed with turmeric powder (*Curcuma longa* L.). It applied on the swellings as well as on the wounds.

IV. Half cup bark extract is mixed in a cup of warm water with some jaggery. Two teaspoon of it thrice a day is advised in the treatment of urinary tract infection.

## 4. ***Abrus precatorius*** **L.**

**Family:** Fabaceae

**Local Name:** Gunj, Ratti

Perennial twiners; leaves long, pinnate, opposite or subopposite, apiculate; flowers pink or white; seeds globose, blood red with the black botch around the hilum or the seeds entirely white.

**Sacred Aspect**

I. Rosary of the seeds tied on the hand if a person obstructed by evil spirit.

II. The tribal beliefs are the seeds are used for black magic in many regions by the tribals.

**Medicinal Uses**

I. Root paste is used as antiseptic on cuts and wounds.

II. A cup of leaf extract is diluted with a glass warm water. Half teaspoon ginger (*Zingiber officinale*) extract and three teaspoon honey is added in this extract. Two teaspoon of this mixture is given thrice a day in the treatment of the cough until its cure.

III. A teaspoon powder of seeds is used once daily for blood purification for a weak.

IV. In gum swelling, fresh leaves are chewed daily for 3-4 days.

**Miscellaneous Uses**

I. Seeds are kept in to the house for to prevent entry of snakes in house.

**C.N.:** Tribal and rural people cultivate around abodes.

## 5. ***Acacia chundra*** **Willd.**

**Family:** Mimosaceae

**Local Name:** Khair

Shrubs or small tree, bark rough and blackish; leaves long, pinnate 10-20 pairs, flowers white or cream coloured; pods stalked, thin brown glabrous.

**Sacred Aspect**

I. The pods are burried around the four corners of the house for preventing the evil eye in Kokani tribe.

II. Tribals celebrate Dindan or Ger festival. They dance during this festival. While dancing they uses 'Tiparya'(sticks) made from the wood of this tree.

**Medicinal Uses**

I. Bark powder is recommended to use as tooth powder in the treatment of toothache.

II. Bark is also used as Kattha powder in chewing betel leaf (*Piper betle* L.)

III. The bark paste is topically used in the treatment of boils and ulcer.

**Miscellaneous Uses**

I. Wood is used for handles of agricultural tools.

## 6. *Acacia ferruginea* DC.

**Family:** Mimosaceae

**Local Name:** Devbhabul

Tall trees, bark rusty brown; pinnae 3-6 pairs, leaflets 10-20 pairs, linear-oblong; rounded at the ends; flowers creamy white, in long axillary spike; pods flat, brown, straight, glabrous, apiculate, reticulate, seeds 4-8.

### Sacred Aspect

I. Wood is used in to create fire (Agni) by Parshi community in some ritual.

II. Inflorescence is used in the worship of Gramdevta by the tribals.

**Medicinal Uses**

I. A cup of decoction of root is mixed with a teaspoon honey and half teaspoon ginger (*Zingiber officinale* Rosc.) extract. Two teaspoon of it thrice a day is prescribed in the treatment of throat infection until cure.

II. Paste of the flowers is rubbed on the head to treat headache.

III. A glass of bark decoction is mixed with three teaspoon honey. A cup of it, twice a day, is recommended in the treatment of malaria until it cures.

**Miscellaneous Uses**

I. Bark is also used as rat and snake repellent.

II. Gum is edible.

## 7. *Acacia nilotica* (L.) Del. ssp. *cupressiformis* (Stewart) Ali and Far.

**Family:** Mimosaceae

**Local Name:** Ramkathi

Tall trees with cupressiform appearance, bark rusty brown; leaves pinnate; fissured pinnae 3-6 pairs; the pods being generally narrowly and regularly constricted between seeds; flowers yellow in heads.

**Sacred Aspect**

I. Wood is used in to create fire (Agni) by Parshi community in rituals.

II. Twigs are used in Havan.

III. It is the believed that the evil spirits stay on the thorny branches of the tree. These people hence stay away from the tree at night.

**Medicinal Uses**

I. A teaspoon bark powder is diluted with a cup of warm water and three –four drops of lemon juice [*Citrus aurantifolia* (Christm.)Sw.]. A cup of this mixture is advised twice in a day in the treatment of diarrhoea.

II. Pods are also used in hair wash to treat hairfall.

III. Leaf paste is applied on the skin to treat ringworm and itching.

IV. Infusion of leaf is used topically in eye problems.

## 8. *Acalypha indica* L.

**Family:** Euphorbiaceae

**Local Name:** Kokali

Erect annual; leaves glabrous, ovate or rhomboid-ovate, acute or obtuse, shortly acuminate, serrate; flowers green, in axillary spik; capsule trilobed, seeds ovoid, smooth, pale brown.

**Sacred Aspect**

I. Occurrence of plant near abodes is thought auspicious.

**Medicinal Uses**

I. A cup of decoction of whole plant is diluted with a cup warm water with two teaspoon sugar. Two teaspoon thrice a day is advised to cure the cough for six days.

II. Leaf paste is topically applied on snake and insect bite.

III. Paste of leaf is locally applied on skin diseases.

## 9. *Achyranthes aspera* L.

**Family:** Amaranthaceae

**Local Name:** Aghada, Chirchir

Erect, woody, tall herbs; leaves elliptic; ovate, membranous, acute or rounded at base; flowers pale green, utricles oblong cylindric, truncate at the apex.

**Sacred Aspect**

I. Vegetable prepared from the leaves is used as the food offering to god.

II. Twenty one plants are carried by the Hindu women while going to river for bath on the day of Rishi-Panchami.

**Medicinal Uses**

I. A teaspoon dried powder is mixed with a glass water. It is prescribed every morning and evening to a patient for kidney stone.

II. Stem juice is used as mouth wash in the treatment of toothache twice a day.

III. Dried leaf powder, about a spoonful, is taken once a day in the treatment of liver problem till cure.

**Miscellaneous Uses**

I. Broom made from stems is used for sweeping.

**C.N.:** It is exotic, native of Tropics.

## 10. *Acorus calamus* L.

**Family:** Araceae

**Local Name:** Vekhand, Yekhand

Perennial herb, rhizomatous, rhizome aromatic; leaves are erect yellowish-green, radical, sheathed at pink base, sword-shaped, flat and narrow, leaf margin; flowers is born in spadix; fruit is a berry.

**Sacred Aspect**

I. Dried plants are tied to the roof of a house for prevention from the evils and bad spirits by Mauchi tribe.

**Medicinal Uses**

I. A cup decoction of the bark is mixed with a cup warm water with two teaspoon honey or jaggery is prescribed two teaspoon thrice a day in the treatment of abdominal pain.

II. Paste of the stem is topically applied on scabies.

III. A cup of bark decoction mixed with sugar or jaggery is prescribed, three teaspoon twice in a day, before sunrise and after sunset in the treatment of malaria.

**Miscellaneous Uses**

I. Dried plants are placed near above the door and windows for preventing the entry of snake.

**CN:** It is planted near abodes.

## 11. *Adansonia digitata* L.

**Family:** Bombacaceae

**Local Name:** Gorakh-chinch

A large tree, the trunk having the large diameter at base and rapidly becoming narrow upwards, bark is gray; leaves digitately compounds, leaflets 3-9, obovate, oblong, acute or obtuse, glabrous, long petioled; flowers white, large, solitary, axillary; drooping on long peduncles; fruits oblong, woody, pulpy.

**Sacred Aspect**

I. It is believed that the tree is having the power to fulfill one's wishes. It is hence called as Kalpavruksh.

II. Fruit is kept hanging in abodes being auspicious.

**C.N.:** It is exotic, native of tropical Africa. It is planted for its curiouse look at common places and gardens.

## 12. *Aegle marmelos* (L.) Corr.

**Family:** Rutaceae

**Local Name:** Bel

Tall trees, armed with thorns, bark greyish white or gray brown, young part pubescent; leaves pinnately 3-foliate; flowers bisexual, greenish or creamy white, scented, in axillary panicles; fruits globose, rind hard grey or yellowish when ripe, pulp sweet, thick, orange coloured, seeds obliquely lenticular or oblong, pale brown.

**Sacred Aspect**

I. Leaves and fruits are commonly used in the worship of God Mahadev Shankar by Hindus.

II. Rural and even urban people belive that Lord Mahadev resides in this tree, hence its leaves are offered religiously during worship.

**Medicinal Uses**

I. Pulp of fruit, about a teaspoon, is given once a day to check dysentery.

II. Leaf juice, about half cup, is taken orally for diabetes twice a day to control sugar.

III. Leaf juice, about half cup, is taken to stop omitting.

IV. Leaf juice, about half cup, is taken once a day for a weak to cure dysentery

**C.N.:** The leaves are stripped off on large scale. Fruits are also collected. This has hampered its natural regeneration. But it is planted at religious places and hence conserved.

## 13. *Ailanthus excelsa* Roxb.

**Family:** Simarubaceae

**Local Name:** Maharuk

Tall tree, deciduous, bark smooth, greenish white, or gray; leaves pinnate, crowded at end of branches; flowers dull yellow, in large, axillary and terminal pubescent penicles; 1 seeded, membranous, prominently nerved, flat.

### Sacred Aspect

I. Twigs of the plant are used for the decoration of the marriage pandal by tribals.

### Medicinal Uses

I. A cup of decoction of stem and bark is mixed with a cup milk. Cup of it is given for checking the excess bleeding and abdominal pain, especially after delivery.
II. Paste of leaves is topically applied to cure ringwarm for few days.
III. A cup decoction of leaves is mixed with ginger (*Zingiber officinale* Rosc.), and sugar. This syrup is prescribed two teaspoon twice a day, in the treatment of cold and cough for ten days.

## 14. *Albizia lebbeck* (L.) Willd.

**Family:** Mimosaceae

**Local Name:** Shirish

Medium sized or tall tree, bark greyish or pale brown; leaves bipinnate, 2-4 pairs, leaflets 4-10 pairs, oblong elliptic, terminal obovate; flowers greenish, fragrant, pale yellow or white, in globose umbellate fragrant heads; pods flat, brown-yellowish.

### Sacred Aspect

I. Inflorescence is used for worship of local deity Gramdevta by Bhil tribe.
II. The tree found planted in many sacred places.

### Medicinal Uses

I. Residue of boiled leaves is applied on the wounds and swellings.
II. Gum powder about half spoon is given once a day to cure ulcer.
III. A glass of bark decoction with two teaspoon honey is advised twice a day for blood purification for a fortnight.

**C.N.:** It is exotic, native of North Australia and Tropical Asia. It is usually planted at public places and along roads for shade.

## 15. *Allium cepa* L.

**Family:** Liliaceae

**Local Name:** Kanda

Perennial, scapigerous herbs with bulbous rootstock, bulb tunicated; leaves fistular; flowers white, regular, acute, rounded umbels at apex of scapes; capsules subglobose, seeds black.

### Sacred Aspect

I. Bulbs are thought preventive for evil eyes.

II. It is used in black magic also by tribal and rural people.

III. Jain and some Hindu communities avoid its consumption during four holy months called Chaturmas.

### Medicinal Uses

I. Constant smelling stops in unconsciousness.

II. Paste of turmeric powder (*Curcuma longa* L.) about teaspoon and a bulb is applied every night on swollen knees until cure.

III. Half cup of decoction of bulb is used in the treatment of asthma once a day for nine weeks.

IV. Spoonful of the extract is dropped in the ears for sunstroke.

### Miscellaneous Uses

I. It is cultivated for edible and spicy bulbs, leaves are used as vegetable.

**C.N.:** It is exotic, native of Western Asia. It is cultivated as a crop in this region.

## 16. *Allium sativum* L.

**Family:** Liliaceae

**Local Name:** Lasun

Plant bulbous, scapigerous, bulb with membranous outer scales in the axils of which are 10-12 fleshy scales; leaves flat, linear, sheaths about half as long as blades; flowers white, in umbels; capsules oblong-ovoid, membranous.

### Sacred Aspect

I. Jain and some Hindu people and also some other tribes prohibit its consumption during four holy months called 'Chaturmas'.

II. It is also used in black magic by tribals.

**Medicinal Uses**

I. A glass of decoction of bulblets is mixed with three teaspoon honey. A teaspoon of it twice a day in the treatment of cough and cold is adviced for a weak.

II. Few drops of decoction of bulblets are also dropped in the ear problems.

III. Juice of bulblets, about teaspoon, is given regularly to increase fertility.

IV. Juice of bulblets, about teaspoon, is given once a day in the treatment of nervous disorder.

**C.N.:** It is exotic, native of Europe, cultivated for its spicy bulblets in this region.

## 17. *Aloe vera* L.

**Family:** Liliaceae

**Local Name:** Korphad

Scapigerous herbs with fibrous roots; leaves ensiform, fleshy, dense, aggregate, pale green with distance horny prickles on the margins; flowers yellow or orange red in racemes; fruits 3-lobed capsules.

**Sacred Aspect**

I. In many houses of Nandurbar district the small twigs is hanged on the roof of house. It is believed that, it's ever greenness is enhances the healthy environment of the house and bring prosperity.

**Medicinal Uses**

I. Latex is used as mosquito repellant in the tribal region.

II. Latex is locally applied regularly on skin for softness and fairness.

III. Latex is recommended for hair wash to treat hairfall.

**C.N.:** It is exotic, native of North America. It is usually planted near temples or abodes for its attractive look and various utilities.

## 18. *Alstonia scholaris* R.Br.

**Family:** Apocynaceae

**Local Name:** Saptaparni

Glabrous laticiferous tree, mature bark is grayish, upper side of the leaves glossy, underside is greyish; leaves in whorls of three to ten, leaves leathery, narrowly obovate to very narrowly spathulate, base cuneate, apex usually rounded; flowers white, in dence cymes and pubescent; seeds oblong, follicles elongated.

**Sacred Aspect**

I. Seven leaved twig is used in the worship of Varundevta in the Toranmal and Dhadgaon region.

**Medicinal Uses**

I. Milky latex of the tree is topically used as an antiseptic on the cuts and wounds.

II. Half cup of root extract is mixed with two cup of boiled water with some of jiggery. It is given two teaspoon twice a day in the treatment of malaria up to ten days.

III. Latex is also used topically in the case of snake bite.

**C.N.:** Recently, it has been planted as a roadside tree on large scale for shade. It is not consumed by animals and hence favoured for plantation.

## 19. Amaranthus hybridus L.

**Family:** Amaranthaceae

**Local Name:** Rajgira

Annual branched herb, leaves long petioled, broadly lanceolate or rhomboid or ovate; flowers minute, reddish or purple; utricles subglobose-ovoid, seeds black.

**Sacred Aspect**

I. Grains are used in the worship of Rhishipanchami by Hindus.

II. It is consumed during fast days.

## 20. *Amaranthus spinosus* L.

**Family:** Amaranthaceae

**Local Name:** Kate – math

Erect, glabrous, annuals, herbs, stem straight usually green, with sharp axillary spines; leaves long petioled, ovate or lanceolate; flowers green, in dense, axillary clusters; seeds dark brown.

**Sacred Aspect**

I. It is auspicious to add leaves in the bath water of the couple that is being married. This is practiced for four days before marriage ceremony.

**Medicinal Uses**

I. Leaf paste is locally applied in the skin problems.

II. A glass of decoction of leaves is added with two teaspoon sugar and a teaspoon lemon (*Citrus aurantifolia* (Christm.)Sw.) juice. it is administered three or twice a day to regulate urination.

III. Paste of stem and leaves is topically used on the wounds of scorpion, snake or any other insect bite.

IV. Half cup of extraction of roots is and two teaspoon honey is administered. Three teaspoon twice in a day is advised for the treatment of fever.

**Miscellaneous Uses**

I. Young leaves are used as vegetable by the tribals.

**C.N.:** It is exotic, native of America, but naturalised on wastelands near human habitations and along roads, rivers *etc.*

## 21. *Ammannia baccifera* L.

**Family:** Lythraceae

**Local Name:** Aagia

Herb, stem erect, green or violet-red; leaves simple, glabrous, often cordate at base; flowers small, in leaf axils; fruit reddish in leaf axils.

**Sacred Aspect**

I. Plant is found near sacred places.

**Medicinal Uses**

I. Leaf paste is topically applied at every night to treat rheumatism until cure.

II. Leaf paste is topically applied to check scabies and ringwarm.

## 22. *Andrographis echioides* L.

**Family:** Acanthaceae

**Local Name:** Kalmegh

Erect herb to a height of 30–110 cm in moist, shady places, stem 4 angled, slender, dark gree;, lance-shaped leaves; small white flowers are borne in spreading racemes; fruit capsule, containing few yellow-brown seeds.

**Sacred Aspect**

I. Plant found at the sacred places.

**Medicinal Uses**

I. A cup of decoction of the leaves mixed with a teaspoon sugar is prescribed twice a day for blood purification and improve the immunity.

II. Half cup of root decoction diluted with a glass warm water. Three to four drops of lemon [*Citrus aurantifolia* (Christm.)Sw.] is added. This is prescribed thrice a day in the treatment of dysentery until cure.

III. A glass of decoction of the leaves and stem is mixed three teaspoon sugar. Three teaspoon of it twice a day in the treatment of fever and malaria is advised.

IV. Paste of the stem and leaf is topically used in case of scorpion bite.

**C.N.:** It mostly grows on calcareous places on old walls of houses and temples.

## 23. *Anethum graveolens* L.

**Family:** Umbelliferae (Apiaceae)

**Local Name:** Balant shepu

Erect, annuals, stem glaucous; leaves long near about 10-20 cm, 3-4 pinnately divided into narrow linear segments; flowers yellow, in large terminal umbels; fruits dorsally compressed, with dorsal ribs thin and sharp, lateral with distinct narrow wings.

### Sacred Aspect

I. It is used as the food offering to goddess Mahalaxmi.

### Medicinal Uses

I. A cup of decoction of whole plant is mixed in a cup of water. Two teaspoon sugar is added in it. Two teaspoon twice in a day is the advised for children for better immunity and stomach problems.

### Miscellaneous Uses

I. Leaves are used as a vegetable.

**C.N.:** It is exotic, native of Europe. It is cultivated for vegetable purpouse.

## 24. *Anisomeles malabarica* R. Br.

**Family:** Labiatae (Lamiaceae)

**Local Name:** Chodhara, Kalatumba

Erect, woody shrubs; leaves opposite, oblong lanceolate, acute to acuminate, crenate- serrate; flowers purple, in dense axillary whorls.

### Sacred Aspect

I. Plant is found at several sacred places. Its occurrence is thought auspicious.

## 25. *Annona reticulata* L.

**Family:** Annonaceae

**Local Name:** Ramphal

Trees with drooping branches; leaves oblong-lanceolate; flowers 2-4 on lateral, sepals minute; fruits yellowish-brown when ripe, seeds black, smooth.

**Sacred Aspect**

I. Goddess Laxmi is worshipped on the day of Laxmipujan. Fruit of this yree are offered to the goddess.

**C.N.:** It is exotic, native of tropical America. This is generally cultivated for edible fruits.

## 26. *Annona squamosa* L.

**Family:** Annonaceae

**Local Name:** Sitaphal

Shrub or small tree, bark is light black, rough; leaves elliptic-lanceolate, simple; flowers solitary or 2-4 on short extra axillary peduncles, green, sepals minute; fruits tubercled, glabrous or cordate ovoid, seeds black.

**Sacred Aspect**

I. Fruit is used as an offering to gods.

II. Found in some sacred places like Patanadevi, Unapdev, Toranmal, Padmalaya, Rail Maruti *etc.*

III. Fruit are used in the worship of goddess Parvati during Hartalika Pooja.

IV. Fruits are used in worship of goddess Laxmi during Dipawali for Laxmipujan.

**Medicinal Uses**

I. A cup of leaf decoction, two teaspoon honey and three drops of ginger juice (*Zingiber officinale* Rosc.) is give twice to thrice a day, in the treatment of dysentery.

II. A cup of decoction of bark prescribed. A teaspoon thrice a day in the treatment of diarrhoea is advised.

III. The paste of the seeds in also used topically on cuts and wounds and also to kill lice.

**C.N.:** It is native of tropical America. It is cultivated for edible fruits but also naturalised near sacred and waste places.

## 27. *Anogeissus latifolia* (Roxb. ex. DC.) Wall.

**Family:** Combretaceae

**Local Name:** Dhavda

Small trees white bark, smooth with rounded patches; leaves elliptic oblong, coriaceous, alternate or sub opposite, obtuse, young leaves silky tomentose, mature leaves glabrous; flowers greenish to pinkish yellow, sessile, in small, dense, pedunculate heads; fruits compressed, orbicular with long beak, wings entire.

**Sacred Aspect**

I. Found in sacred places like Patanadevi, Unapdev, Toranmal, Padmalaya, Manudevi, Kothali *etc.*

**Medicinal Uses**

I. Paste of the bark is topically applied to treat body swellings.

II. Two teaspoon powder of bark is mixed with a cup of water and prescribed as twice a day to check diabetics.

III. Paste of the bark is topically used in the treatment of scorpion bite.

## 28. *Apium graveolens* L.

**Family:** Apiaceae

**Local Name:** Shepu

Erect annuals, stem glaucous; leaves long, 3-4 pinnately divided into narrow linear segments; flowers yellow, in compound umbels; fruits long, dorsally compressed,.

**Sacred Aspect**

I. Dried seeds are used as mouth freshener after the meal at the time of auspicious ceremonies.

**Medicinal Uses**

I. The chewing of seeds is thought beneficial for digestion after the meal.

**C.N.:** It is always cultivated for vegetable.

## 29. *Arachis hypogaea* L.

**Family:**Fabaceae

**Local Name:** Bhuimug

Suberect or decumbent annual; leaves pinnately 4 foliated, obovate-oblong, glabrous or sparsely hairy; flowers yellow or orange, solitary or few in lower axils; pods 1-4 seeded, seeds pinkish or redish, oblong.

**Sacred Aspect**

I. Seeds are generally consumed in the period of fasting.

**Medicinal Uses**

I. Seed oil is used for massage a of child before bath.

II. Seed oil is dropped into the ears of a child for avoiding the dryness.

**C.N.:** It is exotic, native of America. It is cultivated for oil-crop. It is always cultivated for oil seeds.

## 30. *Argemone mexicana* L.

**Family:** Papaveraceae

**Local Name:** Pivla Dhotra

Divaricately branched, prickly, annual, lanticiferous, latex yellow; leaves semi-amplexicaule, spiny; flowers yellow, solitary, terminal, bracts foliaceous; capsules 2-3 cm long, prickly, seeds blackish brown.

**Sacred Aspect**

I. It is used for the worship of god Mahadev.

**Medicinal Uses**

I. Leaf paste is applied locally on the wound of dog bite.

II. Half cup leaf extract is mixed with honey or sugar and spoonful ginger paste (*Zingiber officinale* Rosc.) this is advised twice a day in the treatment of cough.

III. Latex and seed oil is topically used on itching and redness of skin.

IV. Root paste is topically applied on the swelling and wounds.

**C.N.:** It is exotic, native of Mexico. It is common weed near sacred places an in also in cropped fields.

## 31. *Argyreia nervosa* (Burm. f.) Bojer

**Family:** Convolvulaceae

**Local Name:** Weli

Robust woody climber, younger stems densely hairy, growing extensively; leaves heart-shaped large, whitish-silver hairy beneath; flowers large, tubular, pink with darker pink or purple centres, borne in cluster on long stalks; fruit a dry brown capsule, rounded.

**Sacred Aspect**

I. Plant found at the sacred places on hedges.

**Medicinal Uses**

I. Boiled leaves are tied with the help of cloth on swellings and cuts.

II. Half cup of fresh leaf extract is used as mouth wash in ulcer treatment.

III. Root paste is topically applied on dog bite for ten days.

## 32. *Aristolochia bracteolata* Lamk.

**Family:** Aristolochiaceae

**Local Name:** Kidamar

Prostrate to suberect, branched herb; leaves 3-7 cm long broad, cordiform-reniform, glabrous; flowers dark purple, axillary, solitary; capsules 1-2 cm long, oblong-ellipsoid, glabrous, ribbed, seeds obcordate.

**Sacred Aspect**

I. Plant found at many sacred places.

**Medicinal Uses**

I. Root paste is topically used on snake and scorpion bite.

II. Root paste is also useful topically on the skin infection.

## 33. *Asparagus racemosus* Willd.

**Family:** Liliaceae

**Local Name:** Shatavari, Chinyar

Branched, tall, scandent, spinous undershrubs with fasciculated tuberous roots, stem, woody, terete; flowers small, white, in axillary or fascicled, simple or branched racemes; berries globose, scarlet when ripe.

**Sacred Aspect**

I. Plant is found Padmalya temple, Unapdev, Manudevi, Muktainagar temple and many more sacred places.

**Medicinal Uses**

I. Stem is deeped in the coconut oil. This oil is used on the treatment of hair fall regularly in tribal region.

II. Powder of stem is widely used for the nourishment of female fertility.

III. A teaspoon root powder is prescribed with a glass of milk in a day to a mother for improving the immunity and maintains the level of breast milk.

IV. Two to three drops of leaf extract are dropped in nose to stop bleeding.

## 34. *Azadirachta indica* A. Juss.

**Family:**Meliaceae

**Local Name:** Kadu-neem

Tall, semi evergreen tree, bark light black; leaves unipinnate, leaflets 7-17, falcate, ovate-lanceolate, serrate, acute, acuminate; flowers white, in axillary panicles or on old branches; drupes elliptic oblong, glabrous, seed solitary, ellipsoid.

**Sacred Aspect**

I. Healthy twig is used in used for making the Gudhi at the festival Gudhi padava.

II. For the worship of Varun Dev god raining (called Pandav), the leaves are wrapped around the waist of a boy and others dance around him. The boy is called 'Dhondya'. Water is poured on Dhondya and people offer them some food grains.

III. It is believed that the tree always keeps off evil spirit.

IV. Many times the tree is planted near the Muslim Majar and Dargah.

V. Shitala is a local diety. The godess is supposed to free a patient from disease, she is offered dates and leaves of this tree are placed on doors, windows and bed of a patient.

**Medicinal Uses**

I. Young twig is used as tooth brush.

II. Two spoonful of decoction of twig is useful on scabies.

III. Leaf paste is applied on the cuts, wound used as an antiseptic.

IV. Half cup of leaf juice leaf juice is taken once a day for blood purification prescribed for blood purification.

## 35. *Bacopa monnieri* (L.) Wettst.

**Family:** Scrophulariaceae

**Local Name:** Brahmi

Suberect, creeping or prostrate, glabrous, fleshy herb; leaves opposite, sessile, obovate-oblong, fleshy, glabrous; flowers blackish; seeds oblong, pale brown, axillary food capsule.

**Sacred Aspect**

I. Found in most sacred places like Patanadevi, Manudevi and Unapdev.

**Medicinal Uses**

I. Two drops of extract of the plant is dropped into the eyes for cooling effect.

II. Decoction of whole plant is prescribed twice a day in the treatment of cough.

III. Extract of the plant is given regularly thrice a day to the patient suffering from loss of memory.

IV. A cup of extract of leaves is mixed with jaggery and is prescribed twice a day to treat headache.

## 36. *Balanites aegyptiaca* (L.) Del.

**Family:** Balanitaceae

**Local Name:** Hinganbet

Shrub or small tree with sharp and hard thorns; leaves bifoliate, leaflets elliptic oblong leaflets or obovate-oblong, pubescent, concave; flowers greenish in short axillary, gray, hairy, cymes or fascicles; drupes ovoid, woody within, angular, yellowish green when ripe, seeds solitary.

### Sacred Aspect

I. Plant is found at several sacred places.

### Medicinal Uses

I. Root paste is applied on insect bite.
II. Fruit pulp is applied on the pimples.
III. Seed oil is applied on the burn and boils until cure.

## 37. *Baliospermum raziana* Keshav et. Yog.

**Family:** Euphorbiaceae

**Local Name:** Danti

Tall, erect undershrubs; leaves broadly ovate, ovate-lanceolate or elliptic to subpandurate, slightly tomentose, extra floral nectaries along the tips of marginal leaf serrations; flowers yellowish green, monoecious, in axillary or terminal, long peduncled recemes; capsules 3-lobed, pubescent, seeds trigonous, oblong, creamy white, mottled with light black.

### Sacred Aspect

I. Found in most sacred places like Manudevi and Unapdev.

### Medicinal Uses

I. A cup of decoction of leaf mixed with ginger (*Zingiber officinale* Rosc.) is given to treat the asthma and bronchitis. A teaspoon of it twice a day is advised.
II. Half cup leaf extract is administered in the treatment of jaundice.
III. Seed paste is also used in the treatment of scorpion or any insect bite topically.
IV. Leaf paste also used as antiseptic on cut and wounds.
V. Root and leaf paste is also used locally in the treatment of piles.

### Miscellaneous Uses

I. Stem is used as tooth brush in the morning by the tribals.

## 38. *Bambusa arundinacea* (Retz.)Willd.

**Family:** Poaceae

**Local Name:** Kalak, Bambu

A tall erect green bamboo with spines, culms thick and erect, culm sheath triangular with spiny hairs; leaf sheath, glabrous, ligule short, leaves are simple,acute apex, alternate distichous, linear entire margin; flowers In compound panicles, grain oblong or linear oblong, furrowed on a side,.

**Sacred Aspect**

I. Dried bamboos are used to prepare Dholi for carrying dead body.

II. Long erect culms of Bambu are placed in a common place and other fuels are arranged around it on the day of Holi festival by tribals. It is then worshipped and ingnited. The ignition ceremony is called 'Melada'.

**Medicinal Uses**

I. Root paste is topically used in the treatment of skin problems.

**Miscellaneous Uses**

I. It is used for house or hut contstraction in tribal area.

II. Used for making furniture and various artfacts like table, chairs, various containers, pet's abode, brooms *etc.*

## 39. *Barleria lawii* T.Anders.

**Family:** Acanthaceae

**Local Name:** Akhar

Branched, erect, appressed-hairy shrubs; leaves elliptic, lanceolate, acuminate, tapering at the base; flowers white, solitary, corolla pubescent outside; capsules obovoid, compressed, seeds as broad as long, obtuse, silky-hairy.

**Sacred Aspect**

I. It is found in monsoon and winter season at some sacred places.

## 40. *Barleria prionitis* L.

**Family:** Acanthaceae

**Local Name:** Kate Koranti.

Bushy shrub, branches glabrous; leaves elliptic, acute and bristle-tipped; flowers yellow to orange yellow, sessile, solitary, axillary; capsules ovoid, orbicular, flat, hairy.

**Sacred Aspect**

I. Flowers are used for various worship by Hindus.
II. Flowers are used for making rosaries in temples to offer gods.
III. Flowers are sold at many sacred places for worship.

**Medicinal Uses**

I. Two drops of leaf extract are prescribed in ear problems twice a day.
II. Powder of root is used in the treatment of swollen gums.
III. Leaf paste is used on cuts, wound.
IV. Leaf juice a cup twice a day is prescribed in the treatment of fever.
V. Flower paste is also used for skin fairness and softness with milk and turmeric (*Curcuma longa* L.) powder.

**C.N.:** It is exotic, native of tropical Africa (Medakkar and Sharma, 2016). It is generally planted near temple gardens and also in home gardens.

## 41. *Bauhinia racemosa* Lamk.

**Family :** Caesalpiniaceae

**Local Name :** Apta

Small tree, bark brown, branches drooping; leaves bilobed, ovate-orbicular; flowers white or pale yellow, leaf-opposed or terminal racemes; pods falcate, compressed, seeds oblong, compressed.

**Sacred Aspect**

I. The leaves are offered as 'gold' to fellow-men, friends and relatives on the eve of Dasera festival of Hindu.
II. The 'Shami' tree [*Prosopis cineraria* (L.) Druce] is worshiped on the day of Dasara (Vijayadashami) by Hindu. If it is not available, the 'Apta' tree is worshipped in its place.

**Miscellaneous Uses**

I. Leaves constitute good feed for goats.

## 42. *Bauhinia tomentosa* L.

**Family:** Caesalpiniaceae

**Local Name:** Ashundar

Small trees, branches terete; leaves broader than long, glabrous, pubescent beneath; flowers yellow; pods stalked, flat, seeds oblong.

**Sacred Aspect**

I. It is planted at sacred places.

**Medicinal Uses**

I. Paste of dried bark powder is topically applied on ringworm.

**C.N.:** It is exotic, native of southern Africa (Martin *et al.*, 1987).

## 43. *Bauhinia variegata* L.

**Family:** Caesalpiniaceae

**Local Name:** Kanchan

Medium sized tree; leaves ovate or orbicular, bilobed; flowers white to pink, in dense panicle; pods flat, oblong, glabrous.

**Sacred Aspect**

I. Flowers are used in the worship of Lord Mahadeva.
II. Bark of the wood is used in Havan (sacred fire).

**Medicinal Uses**

I. Half teaspoon fruit pulp is used regularly once a day for blood purification.
II. Ash of the root and stem is used to clean teeth in the treatment of toothache.
III. Extraction of roots is used as antiseptic in cuts and wounds.
IV. Fresh root paste is applied on ringwarm.

**Miscellaneous Uses**

I. Timber is used in hut construction in rural region.

**C.N.:** It is planted in temple gardens as an ornamental tree. It is exotic, native of China (Debnath and Debnath, 2017).

## 44. *Benincasa hispida* (Thunb.) Cogn.

**Family:** Cucurbitaceae

**Local Name:** Kohla

Large, monoecious vines, stem softly hairy, diffuse; leaves broadly ovate, cordate, lobes ovate-lanceolate; male flowers solitary, yellow, pistillodes short, female flower yellow, solitary, ovary densely pubescent; fruits more or less not ribbed, seeds compressed, ovoid, white or yellowish, marginate.

**Sacred Aspect**

I. Fruit is used as a symbol of Devil in Navratri, and cut it at last day Navami after the Havan of Navratri as the end of the Devil.

**Medicinal Uses**

I. Glass of fruit juice daily is used to decreases weight and to check obesity.
II. Cup of extract of fruit is mixed with some jaggery. A teaspoon of it is advised twice a day in the treatment of cough and fever.
III. Half cup of fruit extract is consumed daily in the treatment of jaundice.

**Miscellaneous Uses**

I. Fruit have great demand as vegetable.

**C.N.:** It is exotic, native of Java. It is cultivated mostly in tribal areas for edible fruits.

## 45. *Boerhavia repens* L. var. *diffusa* (L.) Hook. f.

**Family:** Nyctaginaceae

**Local Name:** Punarnawa

Diffuse, woody herbs with long fusiform root, stem with pink pigment; leaves at each node in unequal pairs, broadly ovate or ovate oblong, obtuse or rounded at apex; flowers pink, small, subsessile; fruits clavate, clothed with densely stalked glandular hairs.

**Sacred Aspect**

I. Plant is found various sacred places *e.g.* Patanadevi, Manudevi, Unapdev, Valziri.

**Medicinal Uses**

I. A teaspoon plant powder is mixed in a cup warm water. It is advised twice in a day in the treatment of jaundice.
II. Half cup of root extract is diluted with a glass warm water. Two-three teaspoon honey is added. It is prescribed in the treatment of stomach ache.
III. A cup of decoction of root is diluted with two cups of water and few drops of ginger juice (*Zingiber officinale* Rosc.) is advices in the treatment of asthma.

**C.N.:** It is exotic, native of South Africa (Struwig and Siebert, 2013).

## 46. *Bombax ceiba* L.

**Family:** Bombacaceae

**Local Name:** Kate Sawar, Hawar

Large, deciduous trees, grey bark; leaves crowded at the end of branches, petiolate, foliate, lanceolate or elliptic lanceolate; flowers crimsons, sessile; fruit large, seeds numerous, pyriform, packed in white cotton.

**Sacred Aspect**

I. Plant is recorded various sacred places like Unapdev, Padmalaya, Nimdurda and Sonkhamb.

**C.N.:** It is exotic, native of America and Australia (Mukhopadhyay and Chakraverty, 2008).

**Medicinal Uses**

I. Two teaspoon of leaf powder is diluted in a cup warm water. A cup of it thrice a day in the treatment of dysentery is administrated.

II. Bark paste is topically applied on the wounds and cuts as an antiseptic.

III. Leaf powder is topically applied against ringworm and scabies.

IV. Bark paste is applied topically on pimples.

**Miscellaneous Uses**

I. Flowers are used as vegetable.

**C.N.:** It is found near tribal dieties.

## 47. *Borassus flabellifer* L.

**Family:** Arecaceae

**Local Name:** Tad

Robust tree, reach a height of 30 metres, trunk is grey, old leaves remain attached to the trunk for several years before falling clearly; leaves fan-shaped and 3 m (9.8 ft) long; male and female flowers borne on separate plants, male flowers less than 1 cm long and form semi-circular clusters on catkin-like inflorescences, female flowers solitary, upon the surface of the inflorescence axis; fruits fleshy 15–25 cm wide, each containing 1-3 seeds, fruits black-brown with sweet, fibrous pulp each seed enclosed within a woody endocarp

**Sacred Aspect**

I. It is planted various sacred places, as ornamental tree in temple gardens.

**Medicinal Uses**

I. Juice collected from the tree before the sunrise (called Neera) is recommended for good immunity.

**Miscellaneous Uses**

I. After sunrise collected juice is consumed as a beverage called Tadi.

**C.N.:** It is generally planted as ornamental tree. It is exotic, native of tropicalAfrica.

## 48. *Boswellia serrata* Roxb.

**Family:** Burseraceae

**Local Name:** Salayi

Middle sized to tall, deciduous trees, bark thin ash coloured, branches pubescent; leaves imparipinnate, crowded at end of branches, serrate or crenate; flowers in axillary racemes, at end of branches; drupes trigonous, red glabrous.

**Sacred Aspect**

I. The tree is found various sacred places like Toranmal, Unapdev and Dhadgaon.

**Medicinal Uses**

I. Paste of the flowers is used in the treatment of wounds and cuts.

II. Seed powder is applied on the snake and scorpion bite.

III. Seed powder is used for good hair growth.

IV. Ash of the bark is used as toothpaste.

## 49. *Bougainvillea spectabilis* Willd.

**Family:** Nyctaginaceae

**Local Name:** Bogan-wel

Large, pubescent, climbing shrub; leaves ovate-oblong or broadly ovate; flowers purple white, bracts elliptic ovate.

**Sacred Aspect**

I. It is planted near most of the temples for ornamental purpose.

**Miscellaneous Uses**

I. It is a good live hedge plant and also planted around the border of the temples.

**C.N.:** It is exotic, native of Brazil.

## 50. *Bridelia retusa* Sensu Hook.f.

**Family:** Euphorbiaceae

**Local Name:** Asana

Tall tree, bark blackish brown; leaves elliptic oblong, sometimes obovate, obtuse or subacute; flowers creamy white; drupes subglobose, greenish purple when ripe.

**Sacred Aspect**

I. It grows at various sacred places like Manudevi, Toranmal and Unapdev.

**Medicinal Uses**

I. A teaspoon bark powder is mixed with a cup warm milk. It is recommended at morning and evening against arthritis for a fortnight.

II. Bark paste is topically applied on cuts and wounds.

## 51. *Buchanania lanzan* Spreng.

**Family:** Anacardiaceae

**Local Name:** Charoli

Large tree, rough bark, straight trunk; leaves glabrous above; flowers greenish-white, or creamy white; drupes obliquely lentiform, smooth, glabrous, 1 seeded.

**Sacred Aspect**

I. It is recorded Pal and Toranmal region. It is used in the preparation of sweetmeats on auspicious occasion.

**Medicinal Uses**

I. Seed oil is topically used for smoothening skin.

II. Seed are eating as a tonic.

III. Seed paste is applied on the pimples.

**C.N.:** It is generally planted for its fruits particularly in tribal dominated areas.

## 52. *Butea monosperma* (Lam.) Taub.

**Family:** Fabaceae

**Local Name:** Palas, Dhak

Middle sized tree, bark rough, ash coloured; leaves 3 foliate, leaflets rhomboid or broadly obovate; flowers flame-coloured, pods flat, velvety.

**Sacred Aspect**

I. Single plant found in the Sonkhamb village. It is believed that Lord Hanuman is stays inside the tree. So it is much auspicious in the region.

II. Vaghdev festival is celebrated by the tribals. Branch of this tree is placed erect in a place. If a tiger looks at this branch, then it is supposed that the tiger will not harm anybody in this region.

**Medicinal Uses**

I. Paste of stem bark is applied for rheumatism regularly.

II. Floral paste is rubbed on the body for treatment of sunstroke.

III. A cup of floral juice is recommended to cure menorrhagia.

## 53. *Caesalpinia bonduc* (L.) Roxb.

**Family:** Caesalpiniaceae

**Local Name:** Sagar-gota

Extensive, scandent, spiny, shrubs or stragglers, branches pubescent; leaves 30-50 cm long, pinnae 6-8 pairs; flowers pale yellow, pods spiny, oblong elliptic; seeds 1-2 oblong, smooth, lead coloured, polished.

**Sacred Aspect**

I. It is recorded at sacred places *e.g.* Patanadevi, Dandapaneshwar, Kapileshwar, Kondaibari, Rameshwar.

II. Seeds are stored for preventing the entry of snake into the house.

**Medicinal Uses**

I. A cup of root decoction is prescribed a teaspoon thrice a day to control cough and fever until cure.

II. Half cup seed extract is diluted in a cup of water and a teaspoon honey it is given two teaspoon twice a day to check malaria.

III. Seed paste is topically applied against scabies.

IV. Half cup decoction of leaves and a teaspoon honey is prescribed thrice a day in the treatment of throat infection.

**C.N.:** It is planted as hedge near abodes, farmyards and religious places. It is exotic native of North America (Debnath and Debnath, 2017).

## 54. *Caesalpinia pulcherrima* (L.) Sw.

**Family:** Fabaceae

**Local Name:** Puraiful

Tall shrubs, bark gray brown, rough; leaves 10-15 cm long, pinnae 4-9 pairs; flowers red or yellow, in terminal racemes; pods flat, seeds 6-10.

**Sacred Aspect**

I. Plant is planted at many sacred places.

**Medicinal Uses**

I. Seed powder paste is smeared once daily on the ringworm.

**C.N.:** It is exotic, native of South Africa. It is planted in home and temple gardens.

## 55. *Calotropis gigantea* (L.) R. Br.

**Family:** Asclepiadaceae

**Local Name:** Rui, Pandhari mandar, Ruchaki

Stout lanticiferous, shrub or small tree, branches white pubescent; leaves large, thick, obovate-oblong, sessile or shortly petiolate, pubescent; flowers white or purplish, follicles 7-11 cm long, paired, seeds numerous, brown flattened, with brown coma.

**Sacred Aspect**

I. Flower garland of white variety is offered to Lord Hanuman a Hindu god.

II. Before going to marry with widow the bachelors marry with this plant is belief.

III. It is believed that is, the delivered lady spirit stay at night in this plant, so no one should cut the leaves of the plant at night.

**Medicinal Uses**

I. A cup decoction of stem bark is diluted with a cup water with two teaspoon honey It is given three teaspoon twice to thrice in day, for the treatment of diarrhoea.

II. Latex is topically applied on the wounds of snake bite and scorpion bite.

III. Latex is also used as toothpaste in the tooth problem.

IV. A spoonful of latex is also used to stop vomiting.

**C.N.:** It is exotic, native of Africa

## 56. *Calotropis procera* (Ait.) R. Br.

**Family:** Asclepiadaceae

**Local Name:** Rui, Arka mool

Stout shrub or small tree, branches white cottony, pubescence; leaves large, ovate, thick; flowers white with purple tips; follicles in pairs, falcately ellipsoid, seeds numerous, coma silky white.

**Sacred Aspect**

I. After death of the person during Daswa (Ten days) an earthen pot filled with water and garland of flowers tied over it. It is believed that sole comes from flowers and drink the water.

**Medicinal Uses**

I. Latex is applied for removing the pierced thorn from the body.

II. Latex is applied on the swelling.

**C.N.:** It is exotic, native of Africa.

## 57. *Canna indica* L.

**Family:** Cannaceae

**Local Name:** Kardali

Perennial herbs with tuberous root stock; leaves oblong, elliptic or lanceolate; flowers rather distant, 2-togather, erect, petals 2-3 cm long, connate at base, upper staminodes bright red or yellow; fruit capsule, seeds granulate, black.

### Sacred Aspect

I. Flowers are used in the worship of gods and godesses.

II. In a fasting period, the rhizome of the plant is used as auspicious food.

**C.N.:** It is exotic, native of tropical America, it is generally planted as ornamental in the home gardens and temples. It is exotic, native of South America.

## 58. *Cardiospermum halicacabum* L.

**Family:** Sapindaceae

**Local Name:** Kapal phodi

Climbing, glabrous, hairy herb, striate, stem wiry; leaves membranous, biternate; flowers small, white, in umbellate cymes on terminal peduncle; capsules trigonous, seeds 3 globose, smooth, black, arillate.

### Sacred Aspect

I. Plant is found at many sacred places.

### Medicinal Uses

I. Two drops of the extract of whole plant are dropped in ears against earache.

II. Paste of whole plant is topically applied on rheumatic joints.

III. Half cup of stem extract is mixed with a cup boiled water and pinch of sugar. It is recommended a cup thrice a day, in the treatment on fever.

**C.N.:** It is mostly found on the hedges around the temples. It is exotic, native of South America.

## 59. *Carica papaya* L.

**Family:** Caricaceae

**Local Name:** Papai

Dioecious, soft-wooded tree, latex milky; leaves palminerved, large, lobed, elongate, petiol; flowers unisexuals, male flower white in long drooping panicles in axils of upper leaves, female flower is pale yellow born solitary or in cluster, larger than male flowers; berries ovoid-oblong, greenish orange to bright orange when ripe, seeds numerous, black, warty.

**Sacred Aspect**

I. Fruit is used in the worship purpose during engagement for marriages.

II. Fruit is consumed during fasting period.

**Medicinal Uses**

I. Daily fruit consumption is useful for digestive problems.

II. Latex is used to treat the ringworm.

**C.N.:** It is exotic, native of tropical America. It is cultivated in home gardens and also in the grounds of temples.

## 60. *Cascabella thevetia* (L.) Lippold

**Family:** Apocynaceae

**Local Name:** Pivla Kanher

Evergreen much branched, shrub or small tree; leaves linear-lanceolate, glossy green, covered with waxy coating; flowers, long funnel-shaped fragrant yellow; fruit is stony, deep red-black, seed large.

**Sacred Aspect**

I. Flowers are used in various worships in many temples.

II. Flowers and the leaves are also used in temple decoration.

**Medicinal Uses**

I. Seed powder paste is used in the treatment of scabies.

**Miscellaneous Uses**

I. Dry fruits are crushed and powdered. It is sprayed around the farm for avoiding the pest and other unwanted organisms.

**C.N.:** It is exotic, native of tropical America. It is planted on sacred places as an ornamental.

## 61. *Cassia absus* L.

**Family:** Caesalpiniaceae

**Local Name:** Tarhotana-mama

Erect herb, clothed with numerous hairs; leaves 3-5 cm long, ovate-oblong or ovate elliptic; flowers yellow, red; pods linear lanceolate, dark brown, compressed, seeds glossy.

**Sacred Aspect**

I. Plant is found at most of the sacred places.

**Medicinal Uses**

I. Leaves are heated and tied on swelling for a overnight to benefit bone fracture.

II. Heated leaves are tied with turmeric as poultice on the thorn injected portion with the help of cloth for overnight to removing the thorn from the body.

III. Flower paste is topically applied to remove pimples.

## 62. *Cassia auriculata* L.

**Family:** Caesalpiniaceae

**Local Name:** Aawali

Branched bushy shrub; leaves pinnate, 8-12 cm long, leaflets oblong-obovate, stipule auricled; flowers bright yellow, in axillary corymbose racemes; pods flat, seeds many, ovate-oblong.

**Sacred Aspect**

I. Inflorescences are used in the worship of Mhasoba a local diety.

**Medicinal Uses**

I. Roots decoction is given three teaspoon twice a day in the treatment of fever for about a week.

II. Whole plant powder, about a teaspoon, is given twice a day in the treatment of jaundice.

## 63. *Cassia fistula* L.

**Family:** Caesalpiniaceae

**Local Name:** Bahava, Baswan, Girmal

Trees, bark yellowish or greenish; leaves pinnate, 20-40 cm long, leaflets large, ovate or ovate oblong; flowers golden yellow, in large drooping racemes; pods terete, dark blackish brown, seeds many, ovate-ellipsoid.

**Sacred Aspect**

I. Plant is found Patanadevi and padmalaya sacred areas.

**Medicinal Uses**

I. Root paste is effective to check ringworm.

II. Dried leaf powder is used in the treatment of cough and fever. The two teaspoonful powder is mixed in a cup goat milk. It is given a cup twice a day to a patient for a week.

III. Paste of the flower is rubbed on gums to cure toothache.
IV. Flower paste is topically applied on joints in the treatment of rheumatism.
V. Young green branches are used as toothbrush to check foul smell of mouth.

**C.N.:** It is also planted as an ornamental in garden. It is exotic, native of North America (Debnath and Debnath, 2017).

## 64. *Cassia occidentalis* L.

**Family:** Caesalpiniaceae

**Local Name:** Dukkar sheng, Ran-takla

Erect, undershrub; leaves bipinnate, oblong-lanceolate, long-acuminate; flowers orange yellow, in axillary and terminal short corymbose racemes; pods flat, shortly stipitate, linear lanceolate, seeds 20-30, subcylindric-ovoid, greenish brown, smooth, glabrous.

### Sacred Aspect

I. It is found at Patanadevi, Padmalaya and Manudevi sacred area as a common weed.

### Medicinal Uses

I. Half cup of root extract is used in the treatment of malaria. Extract is administrated orally, two teaspoon twice a day, to a patient for a fortnight.
II. A cup of root extract is diluted in a cup warm water. It is used twice in a day in the treatment of diabetics to control sugar.
III. Flower paste is topically applied against scabies.
IV. Flower extract is diluted in water. Two teaspoonful of it is advised against throat complaint.

**C.N.:** It is exotic, native of South America. It is naturalised mostly near human habitation on waste places.

## 65. *Cassia sophera* L.

**Family:** Caesalpiniaceae

**Local Name:** Ran-takla

Glabrous, branched shrub or undershrub, branches are green; leaves bipinnate, 10-20 cm long, leaflets 6-10 pairs; flowers yellow, terminal and axillary racemes; pods linear, oblong, seeds ovoid flat, dark brown.

**Sacred Aspect**

I. Plant is found at most of the sacred places.

**Medicinal Uses**

I. Leaves are heated and tied with turmeric paste (*Curcuma longa* L.) on the thorn injected portion with the help of cloth overnight to remove the thorn from the body.

**Miscellaneous Uses**

I. Young leaves are edible, used in the diet as vegetable

**C.N.:** It is exotic, native of Tropical America (Debnath and Debnath, 2017).

## 66. *Cassia tora* L.

**Family:** Caesalpiniaceae

**Local Name:** Takla, Tarwad, Tarota, Powadya

Suffruticose, glabrous, foetid herbs; leaves dark green, leaflets 3 pairs, orbicular to obovate; flowers yellow; pods subtetragonous, falcate, beaked, seeds 20-30, oblong.

**Sacred Aspect**

I. Tribal communities believed that the white impression on the leaves show the snakes died in past, so before they are going for sowing the seeds in their farm, first they worship of these leaves for the remembrance of Nagdevta.

II. Young fresh leaves are cooked and offered to goddesses Mahalaxmi.

III. Viroba is diety of Dhangar community. This diety is worshiped by offering leaves and flowers by this plant species.

**Medicinal Uses**

I. Powder of leaves is topically used in skin infections.

II. Half cup of seed extract is diluted in a glass water. Three teaspoon honey is added and administrated. It is advised twice a day in the treatment of fever for 7-8 days.

III. Paste of the flower is topically applied on the cuts and wounds.

**Miscellaneous Uses**

I. In tribal region, the young leaves are collected and used as vegetable.

**C.N.:** It is exotic, native of Australia. It is naturalised near human habitations.

## 67. *Catharanthus roseus* (L.) G.Don

**Family:** Apocynaceae

**Local Name:** Sadafuli

Erect bushy, perennial herbs, evergreen shrubs, glabrous; leaves oblong-elliptic or lanceolate, membranous, glossy, obtusely apiculate at apex; flowers rosy-purple or white, solitary or paired, follicles oblong, seeds black, oblong.

**Sacred Aspect**

I. It is planted generally near to temples as an ornamental.

**Medicinal Uses**

I. Three teaspoon extract of whole plant is diluted in a cup of water and a cup of it at every morning and evening is given for to control blood sugar.

**C.N.:** It is exotic, native of West Indies. It is cultivated as an ornamental near temple and abodes.

## 68. *Cajanus cajan* (L.) Mill sp.

**Family:** Papilionaceae

**Local Name:** Tur, Tor

Grey tomentose, branched, undershrubs; leaves 3-foliate, flowers yellow, papilionaceous, flat; pods oblong, seeds 2-4, ovoid-oblong, smooth, reddish.

**Sacred Aspect**

I. Roof prepaired from dried stems is placed on house entry at the time of marriage ceremony to word off evils eye in rural part of Khandesh.

**Miscellaneous Uses**

I. It is a commen pulse used in diet.

II. Stems are pliable and hence used to make pots useful for agricultural works.

III. Partition wall of huts or wall of abodes is made from dried stems.

IV.Leaves are good food for cattle.

## 69. *Celastrus paniculata* Willd.

**Family:** Celatraceae

**Local Name:** Jyotishmati, Malkangani

Scandent shrubs, bark pale brown lenticellate; leaves ovate, elliptic, oblong or orbicular; flowers greenish yellow, dark red or orange when ripe, seeds 4-6, brown, ovoid, arillate.

**Sacred Aspect**

I. It is found at of the sacred places like Padmalaya, Astambha, Hanuman Khora, Manudevi, Patanadev, Uanapdev *etc.*

**Medicinal Uses**

I. Seed powder about half spoon is given orally in digestion problems.
II. Oil obtained from seeds is given a teaspoon daily as brain tonic.
III. Seeds oil is used for massage to reduced body ache.
IV. A cup of extract of seed is diluted with water and jaggery is added. It is prescribed a teaspoon twice to thrice a day to check of fever and cold.

## 70. *Celosia argentea* L.

**Family:** Amranthaceaae

**Local Name:** Kurdu, Kirdu

Erect, much branched, annual, branches grooved; leaves variable in shape, linear lanceolate, lower petioled; flowers white or pink, in dense terminal spikes; seeds lenticular, black, shining.

**Sacred Aspect**

I. In the festival of Mahalaxmi the plant is cooked in the purpose of food offering to god. It is similarly offered to Kanbai, a local diety.

**Medicinal Uses**

I. Whole plant powder is used in the treatment of snake bite.
II. A teaspoonful seed powder is mixed in a glass water and recommend thrice a day to treat to the dysentery and also as blood purifier.

**Miscellaneous Uses**

I. Leaves are used as vegetable in locally.

**C.N.:** It is a common weed of waste land and crop fields. It is exotic, native of Tropical Africa.

## 71. *Centella asiatica* (L.) Urb.

**Family:** Apiaceae

**Local Name:** Brahmi

Creeping herbs, rooting at nodes; leaves thick, broadly orbicular-reniform, petiole sheathing at base; flowers pink or red, in axillary, fasciculate umbels; fruits ovoid.

**Sacred Aspect**

I. Plant is found at many sacred places.

**Medicinal Uses**

I. Plant extract about half cup is advised twice a day for a month in the treatment of arthritis.

II. Leaf juice about two spoonful is given regularly to enhance the memory power.

III. A cup of plant extract with a cup of warm milk and a teaspoon honey is given at night in the treatment of hypertension regularly.

**C.N.:** This species is sometimes mistaken or replaced by another species *viz., Bacopa monnieri* (L.) Wettst. (Scorphulariaceae). It is also called as 'Brahmi'.

## 72. *Chlorophytum borivilianum* Sant. and Fernand.

**Family:** Liliaceae

**Local Name:** Safed Musali

Scapigerous, 15-35 cm tall, glabrous herbs, root tubers sessile; leaves elongate, lorate or ensiform; flowers white, in 3-nate clustered, racemes; capsules obovoid, trigonous, seeds black.

**Sacred Aspect**

I. It is believed that the flowers kept a home brings healthy and prosperous family life.

**Medicinal Uses**

I. Tubers are consumed for good health as tonic.

**Miscellaneous Uses**

I. Leaves are edible as vegetable.

## 73. *Cicer arietinum* L.

**Family:** Fabaceae

**Local Name:** Harbhara

Branched annuals, branches glandular, pubescent; leaves pinnate, 9-15 pairs, subopposite to alternate, serrate; flowers pink-yellow; pods turgid, ovoid-oblong, underground, seeds 1-2, brown, oblong.

**Sacred Aspect**

I. Food prepared from seeds is used as offering to goddess during Mangala Guar.

**Medicinal Uses**

I. Seeds with jaggery are recommended for regular consumption as tonic.

II. Seed paste is applied on face at the night for smoothening and fairness of the skin.

**Miscellaneous Uses**

I. Young leaves are also used in diet

**C.N.:** It is exotic, native of South Europe. It is common crop in the region.

## 74. *Cissampelos pareira* L.

**Family:** Menispermaceae

**Local Name:** Pahadvel, Patha

Slender, dioecious twiners, woody at base; leaves alternate, silky pubescent beneath; flowers minute, greenish yellow, males in axillary panicles, female in pendulous racemes; drupes nearly orbicular or subglobose, red when ripe.

**Sacred Aspect**

I. Climber is found at Patanadevi, Torammal and Manudevi.

II. The dark red fruit is offered to deity of Vanadevta by tribal communities.

**Medicinal Uses**

I. A cup of decoction of roots is prescribed in the treatment of miscarriage and infertility in women.

II. Leaf paste is used in the case of boils and burns. It is topically applied.

III. A cup of decoction of roots mixed with ginger (*Zingiber officinale* Rosc.) and sugar is prescribed thrice a day in the treatment of cough and fever.

IV. Leaf paste is topically applied to reduce inflammation.

V. Root paste is smeared on body places infected by ringworms.

**C.N.:** It is exotic, native of South America.

## 75. *Citrullus lanatus* (Thunb.) Matsura and Nakai

**Family:** Cucurbitaceae

**Local Name:** Tarbuj

Prostrate vines with hispid, fistular stems; leaves hairy, lobed; flowers yellow; fruits rounded hispid when young, red inside when ripe, seeds flat, black-brown.

**Sacred Aspect**

I. A festival of Akhaji or Akshay Trutiya is celebrated remembering forefathers by Hindus. A small earthen pot is filled with water and fruit of this species is kept on its mouth collar. It is then worshipped.

**C.N.:** It is exotic, native of tropical Africa. It is cultivated for edible fruit.

## 76. *Cissus quadrangularis* L.

**Family:** Vitaceae

**Local Name:** Kandvel

Extensive, woody, somewhat succulent, tendrilar climber; leaves 5 foliate, leaflets obovate or ovate rhomboid; flowers greenish white; berries deep purple to almost black when ripe, seeds ellipsoid or subglobose.

**Sacred Aspect**

I. Plant is planted at many sacred places.

**Medicinal Uses**

I. Root paste is applied on cuts and wounds topically.

II. Half cup of decoction of roots is prescribed, teaspoon twice a day, to check dysentery.

III. A cup of decoction of leaves and stem is prescribed three teaspoon twice a day in the treatment of stomach-ache.

IV. Stem extract is applied in the treatment of bone fracture.

V. Half cup decoction of stem is prescribed, teaspoon thrice a day, in the treatment of joint pain.

VI. Stem paste is also applied to treat swellings.

**C.N.:** It is planted in temple gardens.

## 77. *Citrullus colocynthis* (L.)Sch.

**Family:** Cucurbitaceae

**Local Name:** Kadu-Indrayan

Monoecious, creeping, scabrid perennial, with a long tap root; leaves ovate, tendrils simple, rarely bifid; flowers pale or greenish yellow, solitary; fruits globose, slightly depressed atop, seeds flat, pale brown.

**Sacred Aspect**

I. Plant is found at near sacred places

**Medicinal Uses**

I. Root decoction about half cup is given thrice a day for a week as abortifacient.

II. Fruit is consumed during stomach ache.

**Miscellaneous Uses**

I. Seeds are edible.

## 78. *Citrus aurantifolia* (Christm. and Panz.) Sw.

**Family:** Rutaceae

**Local Name:** Limbu

Small, glabrous tree, thorns stout; leaves unifoliate, oblong to elliptic ovate; flowers white, solitary or clustered in the axils; fruits globose, green, yellow when ripe, pulp is pale, acidic.

**Sacred Aspect**

I. Fruit is used in the worship of Yahamogimata is tribal region.

II. Fruit is cut and throw in two directions to remove the effect of evil eye.

III. Fruit is also used in black magic.

IV. Fruits with five fruits of chillie and a black doll are tied in thread and hanged it on the entrance of a house on the day of new moon for preventing evil eye.

V. Mhasoba is a local diety related to agriculture usually situated at banks of rivers and rivulets. Lemon fruits and eggs are offered during worship.

VI. Biroba (Bhairav, Kalbhairav) means the Hindu god Shankar. This god is offered during worship a fruit of lemon, eggs with chilies.

**Medicinal Uses**

I. Half cup of fruit extract is diluted in a glass of water. A pinch of common salt is dissolved in it. It is given to controll dysentery.

II. Fruit juice, about half cup, is given twice a day to check in diarrhoea.

III. Fruit juice about half cup, is taken orally before meal regularly for good digestion.

**C.N.:** It is found under cultivation only.

## 79. *Citrus aurantium* L.

**Family:** Rutaceae

**Local Name:** Santra

Glabrous trees, spines few or wanting; leaves ovate-oblong, shortly or bluntly acuminate; flowers white; fruits globose, generally oblate, green or yellow when ripe, pulp is greenish yellow, sweet.

**Sacred Aspect**

I. Fruit is used in various worships.
II. Fruit is used in food offering to god.
III. Fruit is consumed during fast.

**Medicinal Uses**

I. Three teaspoonful powder of the dried rind is mixed in a cup of water. A cup of it is prescribed thrice a day for weight loss.
II. A cup of decoction of fruit rind is given twice a day in the problem of stomach ache.
III. Half cup of powder of rind is mixed with a cup of water and prescribed in the treatment of indigestion and abdominal pain.
IV. Fruit juice is considered beneficial in case of hypertention.

**C.N.:** It is found under cultivation only. It is exotic, native of South China (Pullaiah and Rao, 2002).

## 80. *Cleome gynandra* L.

**Family:** Capparidaceae

**Local Name:** Pandhri Tilwan

Glandular, pubescent, foetid, annul, stem striate, branched at the base, leaves 3-5 foliolate, flowers white, in terminal racemes, capsules 5-8 cm long, terete, striate, seeds dark brown.

**Sacred Aspect**

I. Plant is found at several sacred places.

**Medicinal Uses**

I. Two drops of leaf extract are dropped in ears to treat earache.
II. Leaf paste is topically applied at every night on rheumatism.
III. Paste of seed powder is topically applied on ringwarm and scabies.
IV. Extract of whole plant is applied on scorpion and insect bites.

**C.N.:** It is exotic, native of Tropical America.

## 81. *Cleome viscosa* L.

**Family:** Capparidaceae

**Local Name:** Pivli-Tilwan

Glandular pubescent, foetid, branched, tall herbs; leaves digitate, 3-5 foliate; flowers yellow, axillary, solitary, growing into a lax receme; capsules 3-5 cm long, terete, seeds dark brown.

**Sacred Aspect**

I. Plant found at many sacred places.

**Medicinal Uses**

I. Leaf paste is applied to cure of boils.
II. Half cup of leaf extract is prescribed two teaspoon thrice in a day for two weeks to a lady to increase fertility.
III. Leaf extract about half cup is taken for a week to treat headache.
IV. Fruit juice is prescribed a cup twice a day to check fever.

**Miscellaneous Uses**

I. Leaves are used as edible vegetable by the tribals.

**C.N.:** It is exotic, native of Tropical America.

## 82. *Clitoria biflora* Dalz.

**Family:** Fabaceae

**Local Name:** Garali

Erect herb, angular stem; leaves imparipinnate, leaflets five, ovate-oblong or elliptic; flowers blue, papilionaceous; pods linear, flat, thinly hairy, mucronate, 5-6 seeded, seeds blackish brown, smooth.

**Sacred Aspect**

I. It is found at Manudevi, Patanadevi, Ranipur, Kalapani and other forest sacred sites.

## 83. *Clitoria ternatea* L.

**Family:** Fabaceae

**Local Name:** Gokarna

Twining, rather woody annuals; leaves pinnate, 5-7 leaflets, subcoriaceous, elliptic-oblong or broadly elliptic; flowers blue or white, solitary, axillary on stout pedicels; pods oblong, slightly curved, beaked.

**Sacred Aspect**

I. Flower is used in the worship of Lord Ganesha and Lord Mahadeva.
II. Flower is used in the worship of goddesses Parvati during Hartalika festival.

**Medicinal Uses**

I. A cup of root decoction is prescribed in the treatment of stomach ache.

II. Leaf paste is applied on swellings.

III. Root decoction about half cup is taken a week for abortion.

**C.N.:** It is exotic, native of Central and South America. It is naturalised along hedges.

## 84. *Coccinia grandis* (L.) Voigt.

**Family:** Cucurbitaceae

**Local Name:** Tondli, Tongal

Perennial scandent or prostrate, dioecious vines, roots tuberous; leaves are wide, 3-5 lobed, bright green above, margins denticulate, tendrils simple; flowers white, solitary, axillary, flowers white solitary, axillary; fruit fusiform-ellipsoid, bright scarlet when fully ripe, seeds oblong, compressed, rounded at the apex, yellowish.

**Sacred Aspect**

I. Vegetable prepared from the fruit is offered to goddess Mahalaxmi.

**Medicinal Uses**

I. Fruit juice about half cup of twice a day, is prescribed in the treatment of cough and cold for 2-3 days.

II. Fruit decoction is used in the treatment of diabetes. A cup of it is advised at morning and evening daily.

III. Root decoction about a cup is prescribed to stop the vomiting.

**Miscellaneous Uses**

I. Fruit is used as vegetable.

**C.N.:** It is common climber on hedges. It has naturalised although it is cultivated for fruit vegetable. It is exotic, native of Africa (Medakkar and Sharma, 2016).

## 85. *Cocculus hirsutus* (L.) Diels.

**Family:** Menispermaceae

**Local Name:** Vasanvel, Vasan

Extensive, much branched, creeper; leaves simple pubescent; white to yellowish flowers; fruits 4 to 8 mm in diameter, purple.

**Sacred Aspect**

I. Flowers are used in the worship of Lord Mahadeva.

II. Plant is found at many sacred places.

**Medicinal Uses**

I. Leaf paste is tropically applied on the cuts and wounds.

II. Leaf extract is dropped in eyes, 2-3 drops thrice in a day in case of eye problems.

III. Leaf extract is mixed with jaggery and a cup of water. A teaspoon of it is advised thrice a day in the treatment of fever until cure.

**Miscellaneous Uses**

I. Leaves are used as edible vegetable.

## 86. *Cocos nucifera* L.

**Family:** Arecaceae

**Local Name:** Naral

Tall, unbranched, erect trees, stout; leaves up to 4 meters (15 feet) long, simple, pinnatisect; flowers greenish; fibrous, stony, white, fleshy.

**Sacred Aspect**

I. Coconut is the most auspicious fruit in Hindu religion. It is used in almost every warship, rites, rituals, ceremonies *etc.*

II. For the success of any work or mission a coconut is broken in front of god or goddess. It is throught auspicious.

III. The engaging bride is offered coconut at the time of engagement.

IV. Leaf of coconut tree is used for decoration in the temples.

V. Leaf is also used in worship of Satyanarayan. This worship also performs after the marriage and in the month of Shravan.

VI. Gulabai is local deity especially of unmarried girls in Khandesh. Gulabai is worshiped by using coconut fruit on a water pot and decorated with basal legumes, rice and arecanuts.

VII. Asara is local deity generally observed neer water reservire, rivers and rivules. Hair of children are presented to her after worshiping Asara with coconut, turmeric, dates and some local food preparations. This act is locally called 'Javul'.

VIII. Ekvira is a goddess the temple at Dhule city situated near the bank of Panjara river. Worship of this goddess is performed by offering coconut, 'Sadi'(Sari) and 'Khan'.

IX. Tribals celebrate Nai-Chari festivals by offering coconut to the diety and foodgrains to the worship.

**Medicinal Uses**

I. Coconut water is consumed during urine problems, and generally to patients of any suffering.

II. Coconut water is also recommended in high fever treatment.

III. Half cup of root decoction is prescribed twice a day in the treatment of dysentery.

IV. Embryo is consumed for the fertility.

**C.N.:** It is said to be native of Cocos Islands and North Andamans. It is planted in home garden and near temple also.

## 87. *Corchorus capsularis* L.

**Family:** Tiliaceae

**Local Name:** Chuncha

Erect, glabrous herbs; leaves ovate, lanceolate, glabrous, serrate; flowers yellow, axillary or extra-axillary, solitary or 2-3 in fascicles; capsules 5-valved, black, seeds wedge shaped, smooth, glabrous, brown or blackish brown.

**Sacred Aspect**

I. Plant is found at many sacred places.

**Medicinal Uses**

I. A cup of leaf extract is recommended thrice a day in the treatment of dysentery until cure.

**C.N.:** It is exotic, native of Pantropics (Singh and Srivastava, 2000).

## 88. *Cordia dichotoma* Forst. f.

**Family:** Boraginaceae.

**Local Name:** Bhokar

Middle sized tree, bark is black ash coloured or blackish brown; leaves large, alternate, coriaceous, obtuse, ovate, elliptic or orbicular; flowers white, polygamous, in axillary, and terminal cymose panicles; drupes pale-orange coloured or pinkish to reddish when ripe, pulpy, globose and apiculate.

**Sacred Aspect**

I. Plant is found at many sacred places.

**Medicinal Uses**

I. Consuming of few fruits is recommended for lung complaints regularly.

II. Fruit and leaf decoction is prescribed in the chest infection until cure. A teaspoon of it twice a day is advised for four days.

III. Fruit pulp is applied topically on scabies.

IV. A cup of bark decoction is prescribed in the treatment of fever and dysentery. A teaspoon of it is advised for few days.

**Miscellaneous Uses**

I. The fruit is edible. Immature fruit is used for pickles.

## 89. *Coriandrum sativum* L.

**Family:** Umbellifereae (Apiaceae)

**Local Name:** Kothimbir, Dhane

Strong smelling annuals, stem glabrous, striate; leaves basal and cauline pinnately or ternatly decompounds, petiole sheathing; flowers white or pink, in terminal compound umbels; fruits subglobose, or ovoid, ribbed, mericarps hollow within.

**Sacred Aspect**

I. During Diwali festival, godess Laxmi of Hindus is worshiped. This worship is called Laxmipujan, a sweetmeal called 'Prashad' prepared from coriander seeds and jaggery is offered to frainds and neighbours.

II. Leaves are added in tasting food recipe 'sagoo' and various preparations.

**Medicinal Uses**

I. Decoction of seed, about half cup, is given twice a day in the treatment of chicken-pox.

II. Decoction of seeds is a good tonic for better health.

**C.N.:** It is exotic, native of South Europe.

## 90. *Costus speciosus* (Koen.) J. E. Sm.

**Family:** Zingiberaceae

**Local Name:** Pev

Erect shrubs, with glabrous stem; leaves subsessile, spiral, oblanceolate; flowers white, in terminal, compact spike; capsule obtusely trigonous, deep-red, seeds greyish-black, smooth with white aril.

**Sacred Aspect**

I. Plant is found at many sacred places.

**Medicinal Uses**

I. Half cup of decoction of root is prescribed twice a day in the treatment of body pain.

II. Stem paste is topically applied in the treatment of wounds caused after burns.

## 91. *Crataeva adansonii* DC.

**Family:** Capparidaceae

**Local Name:** Waiwarna, Varuna

Small tree, height up to 25-30 feet, light brown colour bark, white mark on the branches; leaves 15-18 cm long, rough, leaflets are oval; flowers 2-3 cm in diameter, purple, white and yellowish; fruits 2-3 cm diameter, turns red when ripe.

### Sacred Aspect

I. Leaves are used in the worship of gods and goddesses.

II. Leaves are also used in decoration purposes during auspicious ceremonies.

### Medicinal Uses

I. Decoction of bark is about two spoonfuls, advised to check stomach ache till cure.

II. Bark paste is topically applied on the wounds and cuts as antiseptic.

III. Decoction of bark is prescribed one teaspoon twice a day to control fever.

## 92. *Crinum asiaticum* L.

**Family:** Amaryllidaceae

**Local Name:** Gadhani-Kanda, Gadambi Kanda

Bulbs 5-10 cm in diameter, tunicated, narrowed in to the neck, clothed with old leaf sheaths; leaves lanceolate, shortly acuminate; flowers white, in a simple umbel, perianth tube is 10 cm long; capsules ovoid, smooth, glabrous.

### Sacred Aspect

I. Flowers are used in various worships.

II. Flowers are used for decoration purposes as well.

### Medicinal Uses

I. Decoction of rhizome, about half cup, given once a day for stomach infection.

II. Leaf paste is topically applied on swellings.

III. Root paste is topically smered on the pimples.

**C.N.:** It is planted in gardens for beautiful flowers.

## 93. *Chrysanthemum indicum* L.

**Family:** Asteraceae

**Local Name:** Shevanti

Much branched, perennial herbs; leaves alternate, broadly ovate or ovate-oblong, margin lobed; flowering heads rayed, rays white, dise yellow; achenes obovate.

**Sacred Aspect**

I. Flowers are used in worshiping Mangalagauri and also in Navaratri festival by Hindus.

**C.N.:** It is exotic, native of China and Japan. It is cultivated for its fragrant flowering heads.

## 94. *Cucumis melo* L. var. *melo*

**Family:** Cucurbitaceae

**Local Name:** Kharbuj

Prostrate, monoeciousprostrate climber; leaves broad, suborbicular to subreniform, softly hairy, 5-angled or shallowly 3-7 lobed; flowers upto 2 cm across; berries variable in size and shapes, seeds dirty white.

**Sacred Aspect**

I. Fruit is consumed on the days of fast.

II. Akhaji (Akshaytrutiya) festival is celebrated in this region. Fruit of this plant is placed on a waterpot and worshiped.

**Medicinal Uses**

I. Fruits are consumed for avoiding the sunstroke.

II. Unripe fruit pulp is applied on the face for overnight to increase glowing and smoothness of skin.

## 95. *Cucumis sativus* L.

**Family:** Cucurbitaceae

**Local Name:** Kakdi

Prostrate vines; leaves broadly cordate-ovate, palmately 3-5 lobed; flowers are yellow, monoecious, male fasciculate, female solitary, or fasciculate; fruits oblong, reddish brown when ripe, glabrous.

**Sacred Aspect**

I. Fruit is also offered to goddesses Parvati during Hartalika festival.

II. Tribals celebrate Kakada Navay festival. They worships fruits of this plant and then start consuming them.

**Medicinal Uses**

I. A cup decoction of leaf is mixed in a cup of water and pinch of sugar. A teaspoon of it twice a day is prescribed in the treatment of fever.

II. Slice of fresh fruit is placed on eyes for cooling.

III. Fruit pulp is applied on skin for softness and shining.

**C.N.:** It is cultivated for edible fruits.

## 96. *Cucurbita maxima* Duch. ex Lam.

**Family:** Cucurbitaceae

**Local Name:** Bhopla.

Climbing or trailing, villous vine; leaves suborbicular or reniform, nearly entire or shallowly 5-lobed, strigose –hairy; flowers orange yellow, large, solitary, axillary; berries often globose or oblong, depressed at apex, orange yellow when ripe.

**Sacred Aspect**

I. Vegetable prepared from the fruit is used in food offering to god Mahalaxmi.

**Medicinal Uses**

I. Fruit pulp is topically applied on burns.

II. Paste of fruit is applied on skin to treat ringwarm and scabies.

III. Fruit juice, about a glass, is advised to heart patients for preventing cardiac attack.

**C.N.:** It is cultivated for edible fruits. It is exotic, native of South America (Singh and Nigam, 2017).

## 97. *Curculigo orchioides* Gaertn.

**Family:** Hypoxidaceae

**Local Name:** Kala musli, Taalmuli

Perennial herbs, with elongate underground, stout root stock; leaves radical, plicate, sessile or petiolate, linear-lanceolate, glabrescent; flowers golden yellow, in small racemes; capsules 1.2-1.4 cm long, seeds subglobose or oblong, black, strongly rubose.

**Sacred Aspect**

I. It is found near sacred places, especially in hilly forested areas wherein tribal deities are common.

**Medicinal Uses**

I. Paste of rhizome is applied on the cuts and wounds.

II. A cup of roots decoction is mixed with a cup water, five drops of ginger (*Zingiber officinale* Rosc.) and two teaspoon honey. It is prescribed, two-three teaspoon, in the treatment of fever and cold until cure.

III. A cup of decoction of rhizome is mixed with two-three teaspoon of honey. It is recommended twice a day as a health tonic.

## 98. *Curcuma amada* Roxb.

**Family:** Zingiberaceae

**Local Name:** Ambe halad

Rhizomatous, rhizome branched, with mango flavour; leaves long, oblong, lanceolate, radical leaves sheathed, leaves oblong-lanceolate; flowers white or pale yellow, in spikes; capsules 3-valved, seed ovoid-oblong.

**Sacred Aspect**

I. Rootstock is used for the worship of Marimata a local deity.

**Medicinal Uses**

I. Root paste is topically applied on the rheumatism and swelling.

**C.N.:** It is neither wild nor cultivated in the region but rhizomes are sold markets.

## 99. *Curcuma inodora* Blatt.

**Family:** Zingiberaceae

**Local Name:** Ran halad

Perennial, rhizomatous herb, bearing root tubers, inflorescence on the base of the rhizomes; leaves radical, elliptic-lanceolate, glabrous; flowering axis grows to about 20–30 centimetres tall, crowned with enlarged coloured bracts with pink tips; capsules 3-valved, seeds arillate.

**Sacred Aspect**

I. Paste of rhizome is applied on the body of Bride and Groom, two days prior to the marriage ceremony tribals.

**Medicinal Uses**

I. A cup of root decoction is recommended at morning and evening in the treatment cough.

II. A cup of decoction of rhizome mixed with a glass boiled water and a spoon sugar it is administrated in the problem of body pain.

**Miscellaneous Uses**

I. Shoot is used in diet also.

**C.N.:** It is endemic to Maharashtra.

## 100. *Curcuma longa* L.

**Family:** Zingiberaceae

**Local Name:** Halad

Perrineal herbs, rootstock large, ovoid with sessile cylindric tubers, tubers orange-yellow inside; leaves very large, oblong-lanceolate, tapering to base; flowers yellow in terminal spike.

**Sacred Aspect**

I. It is widely used in the worship of Lord Khandoba, a local deity.

II. It is thought auspicious in marriage. Powder paste is topically applied on the body of bride and bride groom for protect them from evil eye and enhance the glowing of skin.

**Medicinal Uses**

I. A teaspoon powder is mixed with the fresh milk cream and applied on face topically for the fairness.

II. Powder paste is applied on skin diseases, wounds and cuts as an antiseptic.

III. A teaspoon of powder mixed in a glass of milk with a pinch of sugar or drops of honey is recommended a glass every night in the treatment of cough till the cure.

IV. Half cup of leaf decoction is mixed with a cup of water and a pinch of sugar. It is prescribed in the treatment of cold a teaspoon thrice daily.

V. Paste of rhizome is applied topically on insect bite and swellings.

**C.N.:** It is sometimes cultivated in the region but rhizomes are sold in market.

## 101. *Cymbopogon citratus* (DC.) Stapf

**Family:** Poaceae

**Local Name:** Gavti-chaha

Perennials, terrestrial, grass, rhizomes, short and compact, stems terete; leaves mostly cauline, distichous, sheathing at base, leaf blades linear, blades mostly flat, inflorescence terminal, solitary, spike; flowers bisexual, spikelets pedicellate, spikelets sessile or subsessile; fruit caryopsis.

**Sacred Aspect**

I. Oil obtained from leaves is used as perfume on auspicious moments.

**Medicinal Uses**

I. A leaf decoction is recommended a cup twice a day, to treat headache and cold.

**Miscellaneous Uses**

I. Used in tea for special taste and fragrance.

**C.N.:** It is cultivated in home gardens for its aromatic leaves. It is exotic, native of South-East Asia and Australia (Singh *et al.*, 2015).

## 102. *Cymbopogon martinii* (Roxb.) Wats.

**Family:** Poaceae

**Local Name:** Rosha-gavat, Ruicha

Perennial aromatic grass, culms procumbent to erect, rarely branched at base, culms terete, nodes swollen, often waxy; leaves, pale green, linear lanceolate, coarsely scabrid along margin, base cordate.

**Sacred Aspect**

I. It is found on the mountain of Astambha Rhishi near Akkalkuwa. A the time of annual carnival, the participants return home with collection of this grass, and tie it to the door. It is believed that the grass is attracting good things and prevents the bad things entering into the home.

**Medicinal Uses**

I. Leaf juice is used as an antiseptic on the wounds and cuts.

II. Leaf paste is topically applied on rheumatism.

III. Seed oil is topically applied in redening of skin.

**C.N.:** It is wild grass and popular for its aromatic leaves.

## 103. *Cynodon dactylon* (L.) Pers.

**Family:** Poaceae

**Local Name:** Durva

A spreading grass; leaf blades gray-green and short, stems are slightly flattened, often tinged purple in colour; flowering spikes are produced in a cluster of two to six spikes together at terminally.

**Sacred Aspect**

I. Grass is believed sacred to Lord Ganesha and hence 21 culms of it are offered on the day of Ganesh chaturthi festival. It is celebrated an large scale in the entire region. A sweet meal called 'Modak' are offered to Lord Ganesha.

**Medicinal Uses**

I. One and half cup of root decoction mixed with pinch of sugar. It is prescribed in the treatment of dysentery till cure.
II. Root extract is dropped, one to two drops, against eye problems for a week.
III. A cup of leaf decoction is diluted with a glass water and administrated in the treatment of headache.
IV. A cup of decoction of is mixed with two spoonful honey. It is advise, two teaspoon twice a day, as a health tonic.

**C.N.:** It is a common weed of waste places and crop fields. It is exotic, native of Tropical Africa.

## 104. *Cyperus rotundus* L.

**Family:** Cyperaceae

**Local Name:** Nagarmotha

Perennial plant, height up to 140 cm; leaves sprout in ranks of three from the base of the plant, 5–20 cm long, culms triangular; florates small, bisexual; fruit is a three-angled achene.

**Sacred Aspect**

I. Bhaldev (Baldev) is a worship related to agriculture, especially about milching animals. During its rite, 'Lavhale' (*i.e. C. rotundus L.*) sand and few small stones are collected. This plant is grown near houses.

**Medicinal Uses**

I. Root paste is applied on the wounds and cuts topically.
II. A cup of decoction of bulb and a pinch of common salt is recommended in the treatment dysentery.
III. Half cup of decoction of roots is recommended twice a day in the treatment of insect bite.
IV. Root tubers are used as mosquito repellents.
V. Tubers are powdered with 'shikakai' and used for hair wash.

**C.N.:** It is a common weed of marshy waste places and crop fields. It is exotic, native of Tropical Africa (Debnath and Debnath, 2017).

## 105. *Dalbergia sissoo* Roxb.ex.DC.

**Family:** Fabaceae

**Local Name:** Sisam

Trees, branches are drooping; leaves 5-7 foliate, leaflets rhomboid, or elliptic ovate, long acuminate; flowers white or cream coloured, in axillary panicles; pods flat, 1-2 seeded.

**Sacred Aspect**

I. Plant is found at many sacred places.

II. It is generally planted in temple campus for shade.

III. Small auspicious house of gods and goddess as called 'Devhara' from its wood.

**Medicinal Uses**

I. A teaspoon leaf extract is prescribed to treat dysentery until cure.

**Miscellaneous Uses**

I. Foliage is good fodder for domestic animals. It also yields good value timber.

**C.N.:** It is planted near temple and along roadsides.

## 106. *Datura innoxia* Mill.

**Family:** Solanaceae

**Local Name:** Pandhara Dhotra

Erect, branched sufructicose herbs; leaves ovate, pubescent, unequal sided at base, obscurely sinuate-dentate; flowers white, solitary, terminal in the fork of branches; capsules globose, clothed with weak, long spines, septifragally dehiscent, seeds reniform, glabrous, pale to dark brown to blackish.

**Sacred Aspect**

I. Flowers and thorny fruits are used in the worship of Lord Mahadev at Mahashivrati and Hartalika festivals.

**Medicinal Uses**

I. A cup of decoction of leaf is advised two teaspoon thrice a day, in the treatment of asthma.

II. Fruit paste is used on rheumatism topically.

III. Leaf paste is used against ringworm and itchings.

**C.N.:** It is a exotic, native of America.

## 107. *Datura metel* L.

**Family:** Solanaceae

**Local Name:** Kala Dhotra

Erect, glabrous, subglabrous suffructicose herbs, leaves pubescent, broadly ovate-lanceolate, variously deeply toothed; flowers purplish, solitary; capsules nodding, globose, brown when ripe, seeds orbicular, flat, glabrous, smooth.

**Sacred Aspect**

I. It is found at many sacred places.

**Medicinal Uses**

I. Paste of the boiled leaves is applied on the joint pain and muscles pain regularly.
II. Half cup of decoction of root is in the treatment of fever.
III. Paste of fruit and roots is topically applied on the boils, scabies and ringwarm.
IV. Seed oil is applied to reduce tumours in body.

**C.N.:** It is exotic, native of Tropical America.

## 108. *Daucus carota* L.

**Family:** Apiaceae

**Local Name:** Gajar

Annual or biannual herbs, stem 60-90 cm tall, rooth thick; leaves pinnately decompounds, lanceolate, dark green; flowers white or yellowish, large, in compound, terminal, globose umbels; fruit oblong.

**Sacred Aspect**

I. Sweetmeat prepared from carrot root is offered to gods.

**Medicinal Uses**

I. Leaf paste is applied on gums in case of tooth problems.

**Miscellaneous Uses**

I. Taproots are consumed raw and used in various preparation as well.

**C.N.:** It is exotic, native of Europe, North Africa and Temperate Africa.

## 109. *Delonix regia* (Boj. ex Hook.) Raf.

**Family:** Caesalpiniaceae

**Local Name:** Gulmohar

Handsome trees, with an equal spread of crown, bark gray; leaves 30-60 cm long, oblong, elliptic; flowers large, crimsons to scarlet; pods woody, dark brown, flat, seeds oblong, smooth brown or creamy white.

**Sacred Aspect**

I. It is planted in the most of the sacred places and along roadside as an ornamental and shade tree.

**Miscellaneous Uses**

I. Bark paste is placed in corners of abodes as insect repellent.

**C.N.:** It is exotic, native of Madagascar.

## 110. *Desmodium gangeticum* (L.) DC.

**Family:** Fabaceae

**Local Name:** Salvan, Shaalparni

Undershrubs, branched and hairy; leaves unifoliolate, oblong, ovate, entire or repand; flowers clustered, in axdillary and terminal raceme, purple pink; pods linear, flat, falcate, seeds suborbicular, smooth, brown.

**Sacred Aspect**

I. Plant is found at many sacred places.

**Medicinal Uses**

I. A cup of decoction of root and leaves is prescribed a teaspoon thrice a day in to check dysentery.

II. Decoction of roots is recommended a teaspoon twice a day in the treatment of fever.

III. Half cup of extract of root and flowers diluted in a cup and water. It is prescribed three teaspoon twice a day in the treatment of cough until it cures.

IV. Half cup of extract of root is diluted with a cup water and jiggery. This syrup is recommended a teaspoon thrice a day to check asthma.

V. A spoon full of root powder mix with a teaspoon honey and given twice a day for cure ulcer.

## 111. *Diospyros melanoxylon* Roxb.

**Family:** Ebenaceae

**Local Name:** Temburni

Large trees, bark is exfoliating in 4-angular scales, black; leaves ovate-elliptic, elliptic-lanceolate or elliptic oblong, alternate or subopposite, acute at base; flowers white, male flowers in subsessile, drooping, panicled cymes, female flowers solitary; drupes globose or ovoid, glabrous, yellow when ripe, seeds compressed.

**Sacred Aspect**

I. Fruit is used in worship of Gramdevta.

II. It is found at sacred places like Patanadevi, Toranmal and Unapdev sacred forest.

**Medicinal Uses**

I. Pulp of ripe fruits is consumed. It is beneficial to check dysentery and diarrhoea until cure.

II. A cup of decoction of bark is diluted with equal quantity of water. It is recommended a teaspoon twice a day in the treatment of fever until cure.

III. Unripe fruit is used as a antidote on snake bite.

## 112. *Diospyros montana* Roxb.

**Family:** Ebenaceae

**Local Name:** Pali

Middle sized tree, bark gray, smooth; leaves ovate-oblong or elliptic, subacute or acuminate, base acute; flowers greenish-white; fruits globose, reddish brown.

**Sacred Aspect**

I. It is found at sacred places be Toranmal, Manudevi and Dhadgaon.

**Medicinal Uses**

I. Floral paste is topically applied to treat ringworm and insect bites.

## 113. *Diplocyclos palmatus* (L.) Jeffrey

**Family:** Cucurbitaceae

**Local Name:** Shivalingi, Gurul

Monoecious, vines; leaves simple, membranous; flowers cream-coloured or pale yellow, male in fascicles, females in solitary or few together; fruits baccate, globose, with greenish and white patches, red with white lines when ripe, seeds ovate.

**Sacred Aspect**

I. Climber is found at many sacred places.

**Medicinal Uses**

I. Two drops of leaf extract is dropped thrice in a day in case of eye infection until cure.

II. Leaf paste is topically applied on the insect bite.

## 114. *Dolichandrone falcata* (Wall.ex DC.) Seem.

**Family:** Bignoniaceae

**Local Name:** Medshing, Marshing

Deciduous tree, bark is grey; leaves simply pinnate, binnate orbicular or obovate; flowers white, fragrant; capsules flat, falcate, acute, seeds 4 angular, smooth, winged at the two ends.

**Sacred Aspect**

I. Plant is found at many sacred places.
II. White fragment flowers are used in the worship local dieties.

## 115. *Drimia indica* (Roxb.) Jessop

**Family:** Liliaceae

**Local Name:** Ran Kanda

Scapigerous herbs, bulb globose-ovoid, tunicated; leaves flat, starp shaped; flowers appear when plant leafless, flowers purplish-green or brownish purple, in racemes; capsules oblong or ellipsoidal, truncate, seeds flat, winged, black.

**Sacred Aspect**

I. It is found at Manudevi, Patanadevi and some sacred places of Toranmal region.

**Medicinal Uses**

I. Paste of bulb is topically applied on the rheumatism and swellings.
II. Bulb paste is also applied on the cuts and wounds.

## 116. *Echinops echinatus* Roxb.

**Family:** Asteraceae

**Local Name:** Utanti

Much branched, rigid, deep rooted, herb, stem prickly; leaves alternate, lyrate-sinuate, spinescent, white woolly beneath; flowring, heads white, spinous; achenes 6-8 mm long, hairy fusiform.

**Sacred Aspect**

I. Plant is found at many sacred places.

**Medicinal Uses**

I. Half cup of extract of whole plant is prescribed two teaspoon twice in a day in the treatment of malaria fever.
II. Root paste is topically applied on injuries.

**C.N.:** It is exotic, native of Afghanistan.

## 117. *Eclipta prostrata* (L.) L.

**Family:** Asteraceae

**Local Name:** Maka

Erect or prostrate, much branched annuals; leaves opposite, sessile, oblong-lanceolate; flowering heads white, peduncled; achenes obconical or triquetrous, compressed with a narrow wing, deep brown to black.

**Sacred Aspect**

I. Plant is found at many sacred places.

**Medicinal Uses**

I. A cup of extract of the whole plant is prescribed, a teaspoon at the night, against infertility in man.
II. Leaf juice is applied on injuries to avoid infections.
III. A cup of decoction of whole plant mixed with jaggery is recommended, two teaspoon thrice a day, to treat asthma and bronchities.
IV. Decoction of roots about half cup is administratered orally as health tonic.
V. Paste of whole plant is used in the treatment of toothache.

**C.N.:** It is exotic, native of South and Tropical America.

## 118. *Elaeocarpus sphaericus* Gaertn.

**Family:** Elaeaocarpaceae

**Local Name:** Rudraksh

Tall, handsome, evergreen trees, reaching to 53 ft in height; leaves dark green, glossy, lanceolate, old leaves turn red; flowers white-pink, in rows along branchlets, corolla fringed; fruit drupaceous, marble-sized, ovoid, fleshy, flesh green-yellow, seed four, tubercled.

**Sacred Aspect**

I. Dried fruits are very auspicious as per the Hindu religion.
II. It is believed that wearing rosary of dried fruits around the neck is auspicious and keeps healthy.
III. The dried fruits are also very auspicious according to Buddha Dhamma.

**C.N. :** Seeds (5 stones) are available in market for religious purpose. It is not grown in this region.

## 119. *Eleusine coracana* (L.) Gaertn.

**Family:** Poaceae

**Local Name:** Nachni

Annual grass, leaf blades lanceolate, shiny; leaves and culms are typically green in colour, leaf sheaths are prominently flattened; inflorescence spike, spikelets are 5-8 mm long and 3-4 mm wide, grains globose.

**Sacred Aspect**

I. Bread prepared from the seed flour is used in food offering to goddess Kanbai in Khandesh and Dongarydev of Pawara tribe.

**C.N.:** It is exotic, native of Africa (Singh and Nigam, 2017).

## 120. *Emblica officinalis* Gaertn.

**Family:** Euphorbiaceae

**Local Name:** Avala

Tall trees, bark greenish-grey or reddish; leaves distichous, linear, margins glabrous; flowers small, greenish yellow; drupes fleshy, globose, pale greenish yellow.

**Sacred Aspect**

I. Plant is found at many sacred places.

II. During the festival of Aawala Ashtami Hindu people worship Awala tree.

**Medicinal Uses**

I. Dried fruits powder, about spoonful, is taken before meal for digestion.

II. Fruits are consumpted to overcome the gum afflictions.

III. Fruit juice is advised every night, about a glass, to the pregnant lady to increase blood.

IV. Dried fruit paste is applied on the scalp in the case of hair fall and used as natural hair conditioner.

## 121. *Enicostema axillare* (Lam.) Raynal

**Family:** Gentianaceae

**Local Name:** Chota-karait, Nai

Perennial, erect herb, branches many from base; leaves sessile, linear lanceolate or oblanceolate, obtuse; flowers axillary, white; capsules ellipsoid-oblong, glabrous, smooth seeds many, minute, subglobose, brown.

**Sacred Aspect**

I. It is found at many sacred places.

**Medicinal Uses**

I. A teaspoon powder of plant is mixed with honey. It is advised twice a day for blood purification for a month.

II. Paste of the entire plant is used on the insect bites.

## 122. *Euphorbia hirta* L.

**Family:** Euphorbiaceae

**Local Name:** Dudhi

Slender herbs, yellowish crisped hairs; leaves opposite, ovate or ovate-lanceolate, serrate or dentate, dark green or reddish twig above; involucres numerous, globose cymes; capsules appressedly hairy, seeds ovoid and angled.

**Sacred Aspect**

I. Plant is found at many sacred places.

**Medicinal Uses**

I. A cup of decoction of the whole plant is taken orally to treat scabies.

II. A cup of extract of whole plant is prescribed two teaspoon twice a day to control bronchitis and cough.

III. A cup of decoction of whole plant diluted with a glass water and honey is prescribed at morning and evening to treat kidney stone.

**C.N.:** It is exotic, native of Tropical America, but naturalised as a common weed everywere.

## 123. *Evolvulus alsinoides* (L.) L.

**Family:** Convolvulaceae

**Local Name:** Vishnukrant

Trailing or ascending herb with hairy stem; leaves ovate, obovate, elliptic or lanceolate; flowers solitary, blue or blue white; capsules globose, seeds usually 4, oblong, smooth, black or dark brown.

**Sacred Aspect**

I. Plant is found at many sacred places.

**Medicinal Uses**

I. Decoction of whole plant about a cup is taken orally on scorpion bite.

II. A cup of decoction of whole plant is, two teaspoon of it thrice a day, in the treatment of stomach-ache.

III. A cup of decoction of roots and a teaspoon honey is recommended twice a day for the treatment of fever until cure.

IV. Half cup of decoction of leaves and a pinch of sugar is advised in the treatment of asthma.

**C.N.:** It is exotic, native of Amrica (Daniel, 2008).

## 124. *Ficus benghalensis* L.

**Family:** Moraceae

**Local Name:** Vad

Large trees sending down many aerial roots from branches, young part softly pubescent; leaves coriaceous, ovate to elliptic, obtuse, entire; receptacles, sessile in pairs, axillary, globose, red when ripe.

### Sacred Aspect

I. According to Hindu mythology Sati Savitri brought her dead husbands life back from the Yamdev (god of death) under this tree. So during Vat Savitri Pournima women worship this tree to have a long life for her husband.

II. Cutting of the tree is taboo in Hindu religion.

III. It is believed that on the roots live Lord Bramha, on trunk Lord Vishnu and on branches live Lord Shiva.

IV. It is called as Kalpvruksha (wish fullfilling tree).

### Medicinal Uses

I. Root decoction is applied for hair growth.

II. Aerial roots are used as toothbrush to remove foul smell of mouth.

III. Latex is topically used on the skin infections like ringworm.

IV. Leaf paste is used as an antiseptic on cuts and wounds.

## 125. *Ficus racemosa* L.

**Family:** Moraceae

**Local Name:** Umbar, Audumbar

Evergreen tree, bark greyish or blackish brown; leaves ovate, oblong or elliptic to lanceolate, glabrous; receptacle shortly pedunculate, orange red when ripe; achenes reddish brown.

### Sacred Aspect

I. It is believed that Lords Bramha,Vishnu and Mahesh are unitedly called Gurudatta are stay under this tree.

II. Fruit is edible and used in the worship of Gurudatta.

**Medicinal Uses**

I. A teaspoon latex of the plant is mixed with a teaspoon honey and taken orally to check diarrhoea until cure.
II. A cup of fruit decoction diluted with a glass water and a teaspoon sugar is recommended in the treatment of stomach-ache.
III. A cup of root decoction, a teaspoon lemon extract and teaspoon sugar is prescribed in the treatment of dysentery till cure.
IV. Three teaspoonful of fruit extract is taken, twice a day, for controlling the blood sugar in diabetes.

## 126. *Ficus virens* Drynad

**Family:** Moraceae

**Local Name:** Pimpri

Medium sized trees, bark grey; leaves ovate, ovate-lanceolate, acute; receptacles paired, axillary, globose, creamy-white, achenes smooth.

**Sacred Aspect**

I. Pawara people sprinkle water using leaves of this tree in their abodes after their return from cremation ceremony.

## 127. *Ficus religiosa* L.

**Family:** Moraceae

**Local Name:** Pimpal

Large tree, bark is grey or greyish black; leaves coriaceous, ovate-round, entire, acuminate; receptacles in pairs, axillary, sessile, dark reddish-purple when ripe.

**Sacred Aspect**

I. The tree is most auspicious according to Buddhist. Bhagvan Buddha became 'Bodhi' under this tree, hence the tree is also called as 'Bodhivruksh'.
II. In the worship of Bhagwan Buddha, the leaves are commonly used.
III. The people placed the broken and damaged idols and photo frames of gods and goddesses at the base of the tree.
IV. The cutting of the tree is taboo according to Hinduism.
V. Munjoba is the diety which is supposed to residence in this tree. Its temple is generally near this tree. This diety dose not harm if a person circumambulates (Pradakshina) the pimple tree.

**Medicinal Uses**

I. Leaf paste is used as toothpaste in tooth problems.
II. A cup of decoction of leaves and bark is recommended to treat snake and scorpion bite.
III. Three teaspoon latex is mixed with a cup boiled water. A teaspoon of sugar is added. This is prescribed as a health tonic.
IV. Fruits are edible and their consumption enhance digestive capacity.

## 128. *Flacourtia indica* (Burm. f.) Merr.

**Family:** Flacourtiaceae

**Local Name:** Atrun

Small, deciduous thorny tree, bark light blackish-brown, rough; leaves ovate, elliptic or oblong, serrate or crenate; flowers greenish yellow; berries glabrous, deep red or deep purple, seeds obovoid- trigonous, yellowish brown, pubescent.

**Sacred Aspect**

I. Plant is found at several sacred places.

**Medicinal Uses**

I. Four to five fruit daily eaten to cure jaundice
II. Root paste is topically applied on scabies and herpes.
III. Paste of bark powder is topically applied on rheumatic joints.

## 129. *Gloriosa superba* L.

**Family:** Liliaceae

**Local Name:** Kal-lawi

Glabrous, branching, herbaceous climber, tubers plough-shaped; leaves sessile, opposite or in whorls, ovate; flowers yellow, red in upper half, large, showy, axillary, deflexed, solitary or subcorymbose towards end of branches; capsules linear-oblong, seeds subglobose.

**Sacred Aspect**

I. Its flowers are offered to Lord Mahadeva.

**Medicinal Uses**

I. Paste of rhizome is used as antidote on snake bite.
II. Paste of the rhizome is applied topically on the problem of rheumatism.
III. Paste of leaves is applied on the chest, forehead and throat in the treatment of asthma.
IV. Paste of root is topically applied herpes and ringworm.

## 130. *Garuga pinnata* Roxb.

**Family:** Burseraceae

**Local Name:** Kakad, Kakod.

Medium sized, deciduous tree; leaves crowded at ends of branches, compound, leaflets subsessile, elliptic-oblong, ovate-lanceolate; flowers creamy-white, in panicles at ends of branches; drupes subpyriform, fleshy, smooth, black when dry.

**Sacred Aspect**

I. This tree is worshipped by Pawara people on the new moon day.

## 131. *Gmelina arborea* Roxb.

**Family:** Verbenaceae

**Local Name:** Shivan

Deciduous trees, greyish-yellow smooth bark; leaves broadly ovate, acuminate, sub entire; flowers yellow or yellow brown, in terminal or axillary panicles; drupes obovoid, glabrous, yellow or orange when ripe, seeds ellipsoid-oblong, brown and glabrous.

**Sacred Aspect**

I. Plant is found at several sacred places.

**Medicinal Uses**

I. Bark paste is used as antidote for snake poison.

II. Three teaspoon root extract is diluted with a cup water and a teaspoon honey. It is recommended for blood purification for a fortnight regularly.

III. Bark paste is topically applied on bone fracture.

IV. A cup root extract diluted with a glass water. Teaspoonful honey is added in it. This is recommended two teaspoon twice a day, as a health tonic.

**C.N.:** It is exotic, native of Malaya (Madakkar and Sharma, 2016).

## 132. *Gossypium arboreum* L.

**Family:** Malvaceae

**Local Name:** Devkapas

Branched shrubs; leaves palmately 5-7 lobed, coriaceous, subglobose; 2-4 flowers in leaf axils, yellow, white or reddish purple with purple centre; long capsules, pointed, globose, 6-8 seeds per locule, cottony floss white.

**Sacred Aspect**

I. Thread prepared from the cotton fibres obtained the plant is thought sacred. It is hence used in many religious ceremonies and lightning oil lamps and 'munj' in Brahmins.

**C.N.:** It is planted for obtain fibers.

## 133. *Gossypium herbaceum* L.

**Family:** Malvaceae

**Local Name:** Kapas, Kapus.

Large, tall shrub, branches densely hairy; leaves hairy, broadly ovate leaves; flowers yellow with purple centre, solitary, axillary or 2-3 in a cluster; capsules ovoid, acute, cuspidate, seeds light black, reniform, white-cottony.

**Sacred Aspect**

I. Cotton fibres obtained from the plant are used for lighting lamp in various worships.

II. Sacred thread making from cotton called Janhve is worn by a boy in Brahmins.

**Medicinal Uses**

I. Paste of the seed is used in the treatment of cuts and wounds.

II. Seed oil is topically applied on forehead to treat headache.

III. Cake after the oil extraction is good fodder for cattle.

**C.N.:** It is cultivated as agricultural cash crop in the region. It is exotic, native of Arabia and Asia Minor.

## 134. *Haldina cordifolia* (Roxb.) Ridsd.

**Family:** Rubiaceae

**Local Name:** Haldu

Large trees, bark gray-light black, fissured; leaves ovate, suborbicular, cordate; flowers yellow, in globose peduncled heads; capsules obconical, seeds oblong, brown.

**Sacred Aspect**

I. Pawara and Mauchi tribal women worship this tree before conception and wish begetting a daughter and plant the tree after delivery.

II. They also worship this tree if there is no rain or late rain.

**Miscellaneous Uses**

I. Wood is a valuable timber and used in the construction of houses and huts.

## 135. *Hardwickia binata* Roxb.

**Family:** Caesalpiniaceae

**Local Name:** Anjan

Tall tree, bark is dark grey, rough, branches drooping; leaflets paired, sessile, obliquely ovate to rhomboid; flowers yellowish-green, pods strap shaped, veined, glabrous.

**Sacred Aspect**

I. Plant is found at many sacred places.

**Miscellaneous Uses**

I. Foliage constitutes good fodder for milching animals.
II. Wood is valuable timber used for many purposes.

## 136. *Helicteres isora* L.

**Family:** Sterculiaceae

**Local Name:** Murudsheng

Shrubs, stellately hairy, bark smooth, pale or dark brown; leaves bifarious, ovate-oblong or roundish, coriaceous; flowers dark red, in axillary clusters, receptacle raised into gynandrophore; fruits twisted, seeds reddish brown and triangular.

**Sacred Aspect**

I. Plant is found at many sacred places.

**Medicinal Uses**

I. A cup of decoction of stem bark, is prescribed a cup thrice a day to stop dysentery.
II. A cup of fruit extract is diluted with water, ginger and a teaspoon honey is added in it. This is advised two teaspoon twice a day in the treatment of cold until cures.
III. Root paste is applied tropically on cuts and wounds.
IV. Crushed pods are heated with edible oil. They are strained and used as an ear drop also for earache.

**Miscellaneous Uses**

I. Fibers obtained from stem bark is used while constructing hutys and for domestic purpose.

## 137. *Hemidesmus indicus* (L.) R. Br.

**Family:** Asclepiadaceae

**Local Name:** Sariva, Anantamul

Twining or trailing perennials, leaves glabrous, linear to linear-lanceolate, ovate or oblong, flowers few, greenish, axillary, subsessile cymes, fruits follicular, seeds elliptic oblong, compressed.

### Sacred Aspect

I. Plant is found at many sacred places.

II. Root is eaten by tribals to ward off evil spirit.

### Medicinal Uses

I. Half cup of decoction of fruits and roots is prescribed, two teaspoon thrice in a day, in the treatment of fever and cold until cure.

II. A cup of decoction of root is diluted with a cup boiled water. Two teaspoon sugar is added and recommended a for two weeks for blood purification.

III. A cup of decoction of root is administered, two teaspoon twice a day up to ten days to overcome hypertension.

IV. Half cup of extract of roots is recommended a teaspoon at morning and evening as a health tonic.

V. Root juice is used as a cooling drink.

## 138. *Hibiscus rosa-sinensis* L.

**Family:** Malvaceae

**Local Name:** Jasvand

Large shrubs, glabrous; ovate leaves, acuminate, dentate, acute-acuminate, glabrous; flowers solitary on long axillary peduncles, red and showy, stamina tube exerted.

### Sacred Aspect

I. Flowers are offered to Lord Ganesha.

### Medicinal Uses

I. A cup of decoction of flowers is diluted with coconut oil and boiled. This oil is used daily for the treatment of hair loss.

II. Decoction of the flowers also is used as hair conditioner.

III. Root extract is diluted with a cup of water. Teaspoon of sugar is added in it. It is prescribed, two teaspoon twice a day in the treatment of cough.

IV. A cup of decoction of flowers is recommended, a teaspoon twice a day in the treatment of cough until cure.

V. A cup of floral extract is diluted with water and recommended before sunrise with empty stomach for ladies to regularise the menstrual cycle.

**C.N.:** It is exotic, native of China. It is found always under cultivation for beautiful flowers.

## 139. *Hibiscus sabdariffa* L.

**Family:** Malvaceae

**Local Name:** Lal ambadi

Suffruticose, erect, unarmed herbs, with bright to dark reddish purple with glabrous stem and branches; leaves entire, dark-green or reddish-tinged, seldom completely red; flowers solitary, axillary, corolla purple with dark centre; capsules ovoide or subglobose, dark red, seeds large, black brown.

**Sacred Aspect**

I. Plant is found mostly at the sacred places of Toranmal forest.

II. Flowers are used in the worship of forest deity.

**Medicinal Uses**

I. Leaf and fruit is edible.

**Miscellaneous Uses**

I. Fruit pulp is used in diet as like the jam in many parts of Satpura.

II. Calyces are used to sour the foodstuffs and vegetable preparations.

**C.N.:** It is exotic, native of tropical America.

## 140. *Holarrhena pubescens* (Buch.-Ham.) Wall. ex G. Don

**Family:** Apocynaceae

**Local Name:** Dahikudi

Small trees, bark light-blackish brown; leaves ovate or elliptic, glabrous at length; flowers fragrant, white, terminal cymes, follicles elongated, dark green, seeds linear oblong, compressed and brown.

**Sacred Aspect**

I. Plant is found at many sacred places, near the temples or gardens for decorative purpose.

II. Flowers are used for decoration purpose in the auspicious ceremonies and festivals.

**Medicinal Uses**

I. Seed powder is mixed in a glass water and a pinch of sugar is mixed in it. It is advised for better digestion at every night for a month.

II. Three teaspoon of bark powder is diluted in a cup boiled milk. A pinch of turmeric powder and two drops of ginger (*Zingiber officinale* Rosc.) are added in it. This is prescribed, a cup thrice a day in the treatment of malaria until cure.

III. Root and bark paste is topically applied on the rheumatism until cure.

IV. Latex is applied to minimise the trouble caused by herpes.

V. Decoction of roots is administrated to check dysentery. Half cup of it is adjusted daily once for 3-4 days.

## 141. *Holoptelea integrifolia* (Roxb.)Planch.

**Family:** Ulmaceae

**Local Name:** Papda

Large, spreading, glabrous deciduous trees, bark greyish-white or ash coloured, smooth; leaves ovate to abruptly acuminate, rounded subcordate, coriaceous; flowers minute, greenish yellow, samara winged, flat, broadly oblong or orbicular, seeds ovate or ovate-oblong.

**Sacred Aspect**

I. Plant is found at several sacred places.

**Medicinal Uses**

I. Cup of leaf decoction is recommended, a cup twice a day, to reduce body inflammation.

II. Drops of leaf extract are topically applied on scorpion bite.

III. Paste of bark powder is topically applied on the skin fungal diseases.

**Miscellaneous Uses**

I. Seeds are consumed by local inhabitants.

## 142. *Hymenodictyon orixense* (Roxb.) Mabb.

**Family:** Rubiaceae

**Local Name:** Bhorsal

Trees, bark smooth ash coloured; leaves ovate-elliptic or obovate-oblong, pubescent; flowers pale greenish white; capsules stalked, ellipsoidal, brown when ripe, seeds winged.

**Sacred Aspect**

I. Plant is found at several sacred places.

**Medicinal Uses**

I. Drops of root extract are topically applied on snake bite.

II. A teaspoon bark and root powder is diluted in a cup water. It is recommended a cup twice a day to get the relief from fever for five days.

III. A teaspoon bark powder is diluted with a cup goat milk. It is prescribed a cup twice a day in the treatment of malaria fever for a fortnight.

## 143. *Ipomoea nil* (L.) Roth

**Family:** Convolvulaceae

**Local Name:** Nil-pushpa

Herbaceous twiners, stem covered with spreading hairs; leaves entire or more often palmately, 3-lobed; flowers blue turning purple, funnel shaped; capsules broadly ovoid or globose, apiculate, black.

**Sacred Aspect**

I. Plant is found at many sacred places.

**Medicinal Uses**

I. A cup of extract of whole plant is prescribed, a teaspoon at morning and evening as a health tonic.

II. Leaf paste is applied on the cuts and wounds as an antiseptic.

III. Two teaspoon seed powder is diluted with a cup boiled water. It is prescribed, two teaspoon thrice a day, in the treatment of cough and fever.

**C.N.:** It is exotic, native of North America (Patil, 2017).

## 144. *Ipomoea illustris* (Clarke) Prain

**Family:** Convolvulaceae

**Local Name:** Kumrhay

Erect, woody twiners, hairy, pubescent stem; leaves large, pubescent on both sides, ovate to suborbicular, acuminate, apiculate; flowers large, pale violet; capsules enclosed in the calyx, seeds silky-hairy.

**Sacred Aspect**

I. Plant is found at many sacred places.

**Medicinal Uses**

I. Decoction prepared from underground rootstock and a pinch of salt is recommended, a cup thrice a day to check loose motion.

**Miscellaneous Uses**

I. In tribal areas the climber is grown and it spread well on the huts above. It is better for shade and avoids heating of roof.

## 145. *Ixora coccinea* L.

**Family:** Rubiaceae

**Local Name:** Bokadi

Glabrous shrubs, branches angular; leaves sessile, elliptic-oblong or obovate; flowers bright-scarlet; berries globose, purple orange when ripe, glabrous.

**Sacred Aspect**

I. Plant is generally planted near to the temple or in the temple garden for decorative purpose.

II. Flowers are used mostly for decorative purpose or making for the welcome bouquet.

**C.N.:** It is indigenous to India but planted for ornamental purpose. Its scarlet flower attracts attention of passersby.

## 146. *Jasminum officinale* L. var. *officinale*

**Family:** Oleaceae

**Local Name:** Chameli

Twiners or scandent glabrous shrub; leaves opposite, leaflets elliptic or ovate, pubescent; flowers white, fragrant; fruits globose and black.

**Sacred Aspect**

I. Flowers are used in the worship of local deity.

II. Climber is planted generally near the temples.

**Medicinal Uses**

I. Two spoonful of decoction of flower is diluted in half cup of boiled milk or tea and recommended in the treatment of fever for four days.

**Miscellaneous Uses**

I. Flower extract is used as perfume.

**C.N.:** It is exotic, native of China (Patil, 2003).

## 147. *Jasminum sambac* (L.) Ait.

**Family:** Oleaceae

**Local Name:** Mogra

Suberrect shrubs, young branches pubescent; leaves opposite, broadly ovate or elliptic, entire, glabrous; flowers white, in few to many flowered terminal cymes, fragrant, bracts linear-subulate; fruits subglobose, black or dark purple, seeds globose.

**Sacred Aspect**

I. Flowers are used in the worship of local deity.
II. Climber is planted generally near the temples.

**Medicinal Uses**

I. Floral paste is used for skin smoothness.
II. Leaf paste is applied on the cuts and wounds.
III. Floral decoction is used as perfume and room freshener.
IV. Half cup of decoction of flowers, is taken with tea for better digestion.

**C.N.:** It is found exclusively under cultivation for its fragrant flowers. It is exotic, native of Tropical Asia.

## 148. *Justicia adhatoda* L.

**Family:** Acanthaceae

**Local Name:** Adulsa

Erect, branched, glabrous shrub; leaves lanceolate, simple, opposite, ovate-lanceolate; flowers white, irregular, in spikes, in axillary and terminal dense spikes shorter than the leaves; capsules club shaped, subacute, brown.

**Sacred Aspect**

I. Leaves are used in worship of Vanadevta by tribals at Toranmal.

**Medicinal Uses**

I. A cup of decoction of leaves is diluted in a cup of boiled water. Two teaspoon honey is added and prescribed three to four teaspoon of it is prescribed, thrice a day, in the treatment of cough for about a week.
II. A cup of decoction of leaves is diluted with a cup of water and a pinch of sugar is added. Two teaspoon of it is given, thrice a day in the treatment of fever and cold.
III. In the treatment of indigestion, a cup of decoction of fresh leaves is mixed with ginger (*Zingiber officinale* Rosc.) paste. Three teaspoon of it, twice in a day is advice.

## 149. ***Jatropha curcas* L.**

**Family:** Euphorbiaceae

**Local Name:** Moghli errand

Large and soft-wooded shrubs with watery juice, teret branches; leaves, glabrous, broadly ovate to reniform leaves, ocute or acuminate; flowers pale-yellow; capsules globose, yellow and glabrous, seeds oblong.

**Sacred Aspect**

I. Plant is found many sacred places.

**Medicinal Uses**

I. Latex of the plant is topically applied on ringworm.
II. Seed oil is topically applied in the treatment of rheumatism.
III. The twig is used as toothbrush for better oral health.
IV. Latex is topically applied on cuts and wounds as an antiseptic.

**C.N.:** It is exotic, native of tropical America. It is planted as hedge plant around crop field or farmyards.

## 150. ***Kalanchoe pinnata* (Lam.) Pers.**

**Family:** Crassulaceae

**Local Name:** Panphuti

A perennial, succulent and glabrous herb, young part is reddish; leaves decussate, variable, petiolate, crenate or serrate; flowers reddish, pendent, in panicles, seeds numerous, linear.

**Sacred Aspect**

I. Plant is found in gardens at many sacred places.

**Medicinal Uses**

I. Paste of the leaf is applied on the cuts, wounds and boils.
II. Decoction is also applied topically on insect bite.
III. Juice of leaf mixed with a glass water and jaggery is prescribed, a cup twice a day, to reduce the body heat.
IV. A cup of leaf decoction is diluted with on glass water and prescribed, two to three teaspoon thrice a day, to check of dysentery.

**C.N.:** It is exotic, native of Africa. It is always planted for ornamental purpose.

## 151. *Kohautia aspera* (Heyne ex Roth) Bremk.

**Family:** Rubiaceae

**Local Name:** Khetpara

Erect herbs, stem and branches scabrid; leaves sessile, linear-lanceolate, margin revolute; flowers white or very pale blue, subsessile; capsules globose, scabrid with white dots, seeds smooth, glabrous.

### Sacred Aspect

I. Plant is found at several sacred places.

### Medicinal Uses

I. Cup of extract of whole plant is recommended, two teaspoon thrice a day in the treatment of fever for a week.

II. Cup of decoction of whole plant is recommended twice a day in treatment of malaria for 7-8 days.

## 152. *Lagenaria siceraria* (Mol.) Standl.

**Family:** Cucurbitaceae

**Local Name:** Dudhi, Dudhi-bhopala

Large, climbing, monoecious vines, softly pubescent; leaves shortly 3-lobed or angular, petiole 2-glandular; flowers large, white; fruit is greenish yellow, turning brown, seeds white.

### Sacred Aspect

I. Fruit is useful as vegetable in food offering to gods and goddesses locally.

II. On the day of Holi (Melada) festival, tribals tied dried rounded fruits along waist and dence during this festival.

### Medicinal Uses

I. Fruit paste is applied locally on cracks of sole.

II. Fruit juice about half glass, is given regularly to reduce blood pressure.

### Miscellaneous Uses

I. The dried fruits are used as water containers and for musical instrument called 'Pawari'.

**C.N.:** It is exotic, native of Africa (Singh and Nigam, 2017).

## 153. *Lagerstroemia microcarpa* Wight

**Family:** Lythraceae

**Local Name:** Nana Bondara, Hina

Trees, bark ash coloured or greyish-brown; leaves elliptic-lanceolate, acute, glabrous above; flowers white, tinged with pink, in compound panicles; capsules ellipsoid, reddish, seed winged, pale brown and glabrous.

**Sacred Aspect**

I. Plant is found at Patanadevi, Manudevi and many other forest sacred sites.

**Medicinal Uses**

I. A teaspoon bark powder is diluted in a cup of cold water. A pinch of salt is added in it. It is advised in the treatment of stomach-ache until cure.

## 154. *Lantana camara* L.

**Family:** Verbenaceae

**Local Name:** Ghaneri, Gangubai

Straggling shrubs, branches aculeate with recurved prickles, pubescent; leaves ovate to oblong ovate, crenate-dentate; flowers usually orange, bright pink, yellow, red to scarlet; drupes globose, shining, bluish black when ripe.

**Sacred Aspect**

I. Plant is found at many sacred places.

**Medicinal Uses**

I. Leaf paste is applied especially on ringworm disease until cure.
II. Leaf paste is also topically applied on wounds.

**Miscellaneous Uses**

I. It is although a noxious weed. Its stem can be used for making walls of huts and gates of farmyards.

**C.N.:** It is exotic, native of tropical America.

## 155. *Lawsonia inermis* L.

**Family:** Lythraceae

**Local Name:** Kadu mendi, Mehandi

Shrubs, stem with white or brown bark; leaves opposite, entire, broadly lanceolate or obovate-elliptic; flowers small, greenish yellow; capsules globose, dark brown, seeds many, brown.

**Sacred Aspect**

I. Mehandi is having auspicious role in Khandesh region as in Indian cultural life. Leaf paste is applied in design on hand palms and legs as a dye. This is practiced auspicious in ceremonies and festivals.

II. During marriage ceremony a Mehandi ceremony is arranged where paste of is leaves is applied on the hands of bride and bridegroom.

III. Plant is found at many sacred places as hedge.

**Medicinal Uses**

I. Paste is applied a coolant on the head and also used as hair dye.

II. Two teaspoon seed powder is diluted in a glass boiled water. And administrated to overcome to general body weakness.

III. Leaf paste is topically applied on the burns as a coolant.

IV. Decoction of leaf is recommended for gargling in case of throat infection.

**C.N.:** It is exotic, native of North America. It is always planted as hedge around the houses and temples.

## 156. *Leucas cephalotes* (Roxb. ex Roth) Spreng.

**Family:** Lamiaceae

**Local Name:** Ghumma

Erect, hispid herb; leaves broadly ovate, subacute, membranous, crenate-serrate; flowers white, sessile, in axillary whorls, bracts foliaceous, nutlets obovoid-oblong, smooth and brown.

**Sacred Aspect**

I. Plant is found at many sacred places during rainy season

**Medicinal Uses**

I. A cup of leaf decoction is recommended two teaspoon thrice a day, to cure the cough for 7-9 days.

II. Leaf paste is topically applied on cuts and bruises.

III. A teaspoonful whole plant powder is diluted in a cup of cow milk. It is then recommended a cup at every night, as health tonic.

**C.N.:** It is exotic, native of South Africa (Sheikh and Dixit, 2017).

## 157. *Limonia acidissima* L.

**Family:** Rutaceae

**Local Name:** Kavath

Moderate sized tree, with sharp and straight spins, bark greyish; leaves alternate, imparipinnate; flowers small, pale greenish yellow; fruits globose, hard and woody, indehiscent, whitish or pale yellow, seeds many, oblong, compressed and embedded in pulp.

**Sacred Aspect**

I. Cutting of fruit is prohibited at the night because it is believed that the evil spirits rest at night on this tree.

II. Fruit is used in the worship of "Anant puja" in the month of Shravana by the Hindus.

III. Plant is found at many sacred places.

**Medicinal Uses**

I. A cup of root decoction is prescribed, two teaspoon thrice a day in the treatment of body pain until cure.

II. Bark paste is topically applied on cuts, wounds and itching problems.

III. A cup of decoction of fruit is diluted with a cup boiled water. A pinch of common salt is mixed in it. It is recommended, half cup twice a day in the treatment of dysentery.

IV. Consumption of fruit is beneficial in digestion problems.

**C.N.:** It is found under cultivation exclusively.

## 158. *Luffa acutangula* (L.) Roxb.

**Family:** Cucurbitaceae

**Local Name:** Dodke, Shirale

Extensive, scabrous, climbers; leaves broadly ovate, angled or shallowly lobed; flowers pale to bright-yellow; berries linear, pale to dark green, seeds black, flat.

**Sacred Aspect**

I. Plant is found at many sacred places.

**Medicinal Uses**

I. Decoction of tendrils is applied on snake bite.

II. A cup of fruit decoction is prescribed, two teaspoon twice a day, in the treatment of fever until cure.

III. A cup of leaf decoction is recommended, two teaspoon thrice a day, in the treatment of dysentery.

**Miscellaneous Uses**

I. Fruits are edible and used as vegetable in regular diet.

**C.N.:** It is exotic, native of Tropical Asia.

## 159. *Madhuca longifolia* (Koen.) Macbr.

**Family:** Sapotaceae

**Local Name:** Moha, Mahu, Mahuda

Large, deciduous trees, bark is black, greyish black or ash coloured; leaves coriaceous, elliptic-obovate; flowers creamy white, tomentose, axillary cluster at end of branches; berries ovoid, fleshy, yellow when ripe, 1-4 seeded, seeds smooth, shining, brownish black.

**Sacred Aspect**

I. Liquor prepared from its flowers is offered to Gramdevta as worship in tribal region of Toranmal and Dhadgaon.

II. Liquor is served to all the people who are participated in various religious ceremonies.

III. Plant is found at many sacred places.

**Medicinal Uses**

I. Powder of bark is used in the treatment of toothache.

II. Seed oil is used in chest pain topically and also on rheumatism.

III. Ash of leaves is mixed with coconut oil and used on cuts wounds and other skin diseases.

IV. Decoction of flowers is prescribed a teaspoon thrice a day in the treatment of cough and bronchitis.

V. Leaf paste is applied on burns.

**Miscellaneous Uses**

I. Bark powder is used topically on rheumatism.

II. Leaves also constitute good fodder for cattle.

III. Wood is used as fuel.

**C.N.:** All parts of this tree are useful to tribal communities. It is really 'Kalpvruksha' for them.

## 160. *Mallotus philippensis* (Lamk.) Muell.-Arg.

**Family:** Euphorbiaceae

**Local Name:** Shendri

Much branched tree, bark is grey; leaves coriaceous, pubescent, ovate-lanceolate, acuminate; flowers small, males clustered in erect terminal spikes, female solitary, in short spikes; capsule 3-lobed and 3-valved, covered with bright red stellate hairs, seeds glabrous, finely reticulate.

**Sacred Aspect**

I. Dye obtained from the fruit hairs is used as 'Kumku' or 'Kumkum' on forehead by married women.

**Medicinal Uses**

I. A cup of fruit extract is prescribed, two teaspoon thrice a day, in the treatment of dysentery until cure.

II. Seed paste is topically applied on various herpes and ringworm and Seed paste is applied in rheumatic pains.

III. Decoction of seed is also used topically on snake bite.

IV. Fruit extract is applied on boils and cuts.

## 161. *Mangifera indica* L.

**Family:** Anacardiaceae

**Local Name:** Amba

Tall trees with spreading branches, bark is light black or dark brown, inflorescence is pubescent; leaves oblong-lanceolate; flowers cream to yellowish-green, fragrant, in raceme; drupes large, pyriform to subovoid, fibrous, hard, yellow to reddish yellow when ripe, seeds oblong, glabrous, smooth.

**Sacred Aspect**

I. Wreaths of leaves are used in every auspicious festival and ceremony.

II. Fruit is used in worship of Vat Savitri puja.

III. Fruit juice is used for food offering to god as Naivaidya.

IV. Twigs are used for making the wedding hall in marriage ceremony of tribals.

V. Twigs are placed in windows or doors at the time of marriage ceremony. Being auspicious, this is also thought a sign of invitation to the house owners.

VI. Trees are planted most of the time in temple area.

VII. The raw fruits are hanged before the idol of Gaurai, a local deity in khandesh. Gaurai is worshiped on the day of Akshay trutiya (Aakhaji) in summer season. Leaves are also placed before the idol.

VIII. Fruit juice is an important in auspicious ceremonies or festivals.

**Medicinal Uses**

I. A teaspoon of dried gum powder is mixed in a cup of water. It is administrated in the treatment of the toothache.

## 162. *Martynia annua* L.

**Family:** Martyniaceae

**Local Name:** Winchu

Glandular hairy, tall herbs, stem stout, suberect; leaves broadly ovate, lobed, repand dentate; flowers pink or pinkish-white, drooping; fruits woody, glandular-pubescent when young, hooked at tip, seeds black, testa thin.

**Sacred Aspect**

I. Plant is found at many sacred places.

II. Bhil people believed that the fruits if placed on the house roof they prevent the evil spirit.

**Medicinal Uses**

I. Fruit pulp is topically applied on rheumatism.

II. Leaf paste is applied on cuts and wounds topically.

III. Seed oil is applied to stop the hairfall.

**C.N:** It is a weed, it is exotic, native of Mexico.

## 163. *Melia azedarach* L.

**Family:** Meliaceae

**Local Name:** Bakana neem

Moderate sized trees, bark is dark brown, longitudinally furrowed; leaves 1-3 pinnate, ovate-lanceolate or elliptic-lanceolate, acuminate; flowers white with purple tinge, fragrant; drupes ellipsoid-oblong, seeds 4-5, yellow when ripe.

**Sacred Aspect**

I. Plant is found at many sacred places.

**Medicinal Uses**

I. Leaf extract is sprayed as insecticide around abodes.

II. Leaf paste is applied on cuts and wounds as an antiseptic.

III. A cup of leaf decoction mixed with a glass water is prescribed for blood purification.

**C.N.:** It is exotic, native of Myanmar. It is generally planted, as like the 'Neem' tree.

## 164. *Melilotus alba* Medik. ex Desr.

**Family:** Fabaceae

**Local Name:** Ran methi

Erect, glabrous, branched herb; leaves pinnate, leaflets glabrous, oblanceolate, obovate-elliptic; flowers white, small, in axillary, slender racemes; pods turgid, glabrous, seeds globose, smooth, glabrous.

### Sacred Aspect

I. Mauchi people prepared vegetable from the leaves and offer to gods.
II. Plant is found at several sacred places.

**C.N.:** It is exotic, native of Europe.

## 165. *Mimosa hamata* Willd.

**Family:** Mimosaceae

**Local Name:** Arati

Shrubs, stem prickly, glabrescent; branchlets usually red; leaves bippinate, 3-6 pairs, leaflets opposite; flowers pink, globose; pods flat, falcate, 1-seeded joints, seeds joints.

### Sacred Aspect

I. A deity is worshipped on the 5th day after childbirth by Hindus. It is believed that the deity presides over the fate of a new born child. The deity is worshipped using twigs of this plant species along with others such as (i) *Ziziphus maurtiana* Lamk. (ii) *Pennisetum americanum* L. and (iii) *Hibiscus cannabinus* L.

**C.N.:** It is exotic, native of South America (Pullaiah and Ramamurthy, 2001).

## 166. *Mitragyna parvifolia* (Roxb.) Korth

**Family:** Rubiaceae

**Local Name :** Kalam, Kayam

Large trees; leaves elliptic-obovate or suborbicular, cordate, acute at apex; flowers creamy-white, in globose heads; capsules oblong, seeds black, minute.

### Sacred Aspect

I. Indal is festival celebrated by tribals. God Indra is their god. A branch of this tree is placed in a common place. It is then worshipped by offering wine, sal (husked rice grains), fowls and goats. The tribals enjoy dinner after this worship.

### Miscellaneous Uses

I. The timber is used to construct abodes and for agricultural implements.

## 167. *Mukia maderaspatana* L.

**Family:** Cucurbitaceae

**Local Name:** Ankhphod

Scabridly hairy, monoecious climbers; leaves entire, ovate-triangular; flowers pale to bright yellow, male and female in same axil, male pedicellate, female sessile; berries broad and spherical, glabrous and smooth, yellowish-reddish when ripe, seeds grey or light black, compressed.

**Sacred Aspect**

I. Plant is found at several sacred places.

**Medicinal Uses**

I. Root paste is used as toothpaste to get relief from toothache. It is advised twice a day for five days.

**Miscellaneous Uses**

I. Fruits are edible.

## 168. *Merremia dissecta* (Jacq.) Hallier f.

**Family:** Convolvulaceae

**Local Name:** Nil

Twining herbs, stem glabrous; leaves palmately divided almost to the base into 7-9 lobes, glabrous; flowers white, with purple centre, solitary, axillary, rarely in cymes; capsules depressed globose, seeds black glabrous.

**Sacred Aspect**

I. Climber is found at several sacred places.

**Miscellaneous Uses**

I. Plant is recommended to grow around the houses as snake repelant.

**C.N.:** It is exotic, native of America (Patil, 2003).

## 169. *Michelia champaca* L.

**Family:** Magnoliaceae.

**Local Name:** Sonchapha

Tall evergreen tree, bark greyish-brown; leaves ovate to lanceolate, coriaceous; flowers yellow or orange, fragrant; capsules orbicular, seeds 2, scarlet, rounded on black.

**Sacred Aspect**

I. Flowers are used in worship of Hindu Gods *viz.*, Lord Mahadev and Lord Vishnu.

II. Flowers are used as garlands purposes on auspicious ceremony by Hindu women.

III. Perfume of flower is used for fragrance in temple garden.

**Medicinal Uses**

I. Oil obtained from flowers is used for rheumatic joint pains.

II. Leaf paste is topically applied on cuts and wounds.

**C.N.:** It is found under cultivation in home gardens or near temples.

## 170. *Mimusops elengi* L.

**Family:** Sapotaceae.

**Local Name:** Bakul

Evergreen tree, bark grey, sap milky; leaves elliptic ovate, acute-acuminate; flowers dull white, sweet scented; berries ovoid, yellow when ripe, seeds single, black.

**Sacred Aspect**

I. Flowers are used for garlands during auspicious ceremonies.

**Medicinal Uses**

I. Paste of bark powder is topically applied on rheumatism, cuts and wounds.

II. Bark paste is mixed with a teaspoon honey. This is applied on ulcer until cure.

**Miscellaneous Use**

I. Extract of the fruit is used as a tonic.

**C.N.:** It is found only under cultivation in this region, although native of India.

## 171. *Mirabilis jalapa* L.

**Family:** Nyctaginaceae

**Local Name:** Gulbakshi

Large, herbaceous, branched perennials; leaves membranous, ovate, acuminate; flowers solitary, subsessile, perianth white, red, yellow or striped, anthocarps globos, black, ribbed.

**Sacred Aspect**

I. Flowers are used in various worships.

II. Flowers are used for the decoration purposes on some auspicious occagions.

**Medicinal Uses**

I. Leaf paste is applied on cuts and boils.

II. Leaf paste is topically used to reduce inflammation.

**Miscellaneous Uses**

I. Root tubers are used as edible vegetable in diet.

**C.N.:** It is exotic, native of South America. It is planted for ornamental purpose. It has also naturalised fast on waste lands.

## 172. *Momordica dioica* Roxb. ex Willd.

**Family:** Cucurbitaceae

**Local Name:** Kartoli, Katurle

Dioecious tuberous,vines; leaves ovate, tendrils simple, striate; flowers pale yellow, male bracteates, female ebracteate or nearly so; fruits ellipsoid or ovoid, echinate, yellowish red when ripe, seeds variable in size and shapes, embedded in red carmine pulp.

**Sacred Aspect**

I. Vegetable prepared from the fruits is offered to gods.

**Medicinal Uses**

I. Root decoction is diluted with a cup boiled water and two teaspoon honey. It is advised, two teaspoon thrice a day, in the treatment of fever until cure.

II. Fruit paste is topically applied on the swellings for relief.

**Miscellaneous Uses**

I. Fruit is used in regular diet as a vegetable.

## 173. *Moringa oleifera* Lamk.

**Family:** Moringaceae

**Local Name:** Shevga.

Middle-sized soft wooded trees, bark corky, rough gray-brown; leaves usually 3 pinnate, obovate, elliptic-obovate; flowers white, pubescent; capsules more or less 3-angled, beaked, seeds many, glabrous.

**Sacred Aspect**

I. Vegetable prepared from the fruit is used in food offering to goddess 'Mahalaxmi'.

**Medicinal Uses**

I. Syrup is prepared from decoction root two cups of boiled water. A pinch of salt and two tea spoonful jaggery. It is prescribed in the treatment of asthma until cure.

II. Bark powder is used as toothpaste in the toothache problem.

III. Root paste is topically applied on rheumatism and swellings.
IV. A cup of decoction of bark powder is diluted with a cup water. It is prescribed three teaspoon thrice a day for the treatment on stomach ache.
V. Bark paste mixed with gum is rubbed on forehead to check headache.

**Miscellaneous Uses**

I. Pods are used as vegetable in regular diet.

**C.N.:** It is cultivated exclusively for edible fruits. It is exotic, native of America (Singh and Srivastava, 2000).

## 174. *Mucuna pruriens* (L.) DC.

**Family:** Fabaceae

**Local Name:** Kahj-kuili

Large, twining shrubs, hairy when young; leaves 3-foliate, ovate, hairy beneath; flowers papilionaceae, dark purple, pods S-shaped, densely covered with irritant bristles, seeds 4-6, elliptic, compressed, dark brown, spotted black, smooth, glabrous.

**Sacred Aspect**

I. Plant is found at many sacred places.

**Medicinal Uses**

I. Leaf paste is applied on boils.
II. A cup of root decoction is prescribed, twice a day, in the treatment of cough until cure.
III. Seed powder is topically used on scorpion and snake bite.
IV. Stem extract is dropped in eyes about two drops in eye problems.
V. A cup of decoction of roots is diluted in three cup is of water. Juice of lemen [*Citrus aurantifolia* (Christm.)Sw.] and a pinch of salt are added in it prescribed. This is to check dysentery.

## 175. *Musa paradisiaca* L.

**Family:** Musaceae

**Local Name:** Kel

Tree like, stoloniferous stout herbs; leaves oblong, 1-2 m long, pendulous inflorescence, pistillate flowers in lower bract, staminate in upper ones; berries oblong, yellow when ripe, pulp sweet.

**Sacred Aspect**

I. Leaves are used as a dinning plate while offering food to gods and goddesses.
II. Sprouts of the plant are used during the worship of god's *viz.*, Satyanarayana, Ganapati at Ganesh Chaturhi festival, goddess Durgamata at Navaratri festival.
III. Whole plant is though auspicious and hence used by Hindu in decorating pandals for religious ceremonies.
IV. Fruits of the plant are used as the most important gradient of worship.
V. Fruit is consumed as fast food.
VI. Flowers are used in the worship of God Mahadev in Mahashivratri and Hartalika festival

**Medicinal Uses**

I. Leaf paste is topically applied on the treatment of boils.
II. A cup of decoction of root is prescribed a cup thrice a day in the treatment of dysentery.
III. Eating of the fruit is recommended in case of dysentery to avoid the dehydration.

**Miscellaneous Uses**

I. Rhizome, leaves; inflorescence are useful as fodder of domestic cattle.

## 176. *Nelumbo nucifera* Gaertn.

**Family:** Nelumbonaceae

**Local Name:** Kamal

Large, handsome, perennial, aquatic, rhizomatous herbs; leaves orbicular, glaucous; flowers solitary, white or light pink, seeds ovoid and black.

**Sacred Aspect**

I. Flower is thought abode of Laxmi, a godess of wealth.
II. Flower used in the worship of Laxmi at Laxmipujan during Diwali festival.

**Medicinal Uses**

I. A cup of decoction of seeds and two teaspoon honey is prescribed at morning and evening as a health tonic.
II. Stalks of the flowers are used as salad or vegetable.

**Miscellaneous Uses**

I. Thalamus is used as a vegetable in the diet.

**C.N.:** These are growing in ponds near Padmalaya temple, at Padmalay, district Jalgaon.

## 177. ***Nerium indicum*** **Mill.**

**Family:** Apocynaceae

**Local Name:** Kanher

Evergreen shrubs, branches whorled; leaves whorled, elliptic-lanceolate, sparsely hairy beneath; flowers rosy, pink, red or white, follicle lanceolate, seeds villous with terminal coma.

### Sacred Aspect

I. Flowers are used in the worship 'Kanbai' a local diety, especially in Khandesh region of Maharashtra. Worship is started on Tuesday or Friday. The goddess is coconut (a symbol of the sun) on Sunday. The plant 'Kanher' is also regarded a symbol of Lord Krishna.

II. Flower is used in the worship of Gramdevta in many tribal regions.

III. Plant found at many sacred places and planted near the temples.

IV. Flowers are used for decoration.

### Medicinal Uses

I. Latex is topically used on the treatment on boils.

II. Leaf paste is applied on swellings.

III. Extract of flowers is used as a eye drop in eye infection.

**C.N.:** It is exotic, native of Mediterranean region.It is always found cultivated near temple and home gardens.

## 178. ***Nyctanthes arbour-tristis*** **L.**

**Family:** Oleaceae

**Local Name:** Parijat

Large shrub or small trees, rough all over with stiff whitish hairs, branchlets quadrangular, hairy; leaves opposite, ovate; flovers fragrant, white with orange red corolla tube; capsules obcordate or nearly orbicular, compressed.

### Sacred Aspect

I. Flowers are used in various worship.

II. Flowers are used in decoration in auspicious ceremony, marriages and festivals.

III. Flowers are also used as flower garland for women in festivals and auspicious ceremonies.

### Medicinal Uses

I. A cup of decoction of leaves is prescribed, two teaspoon twice a day, in the treatment of cold.

II. Half cup of extract of leaves is diluted with two cup boiled water. Two spoonful honey and ginger paste is mixed in it. It is Prescribed, a teaspoon thrice a day, in the treatment of cough till cure.

III. A teaspoonful leaf powder is diluted in a cup boiled water. It is recommended, two teaspoon thrice a day, in the treatment of fever until cure.

IV. Leaf paste is topically applied on swelling and rheumatism.

V. Paste of the flowers is used to treat burns and skin redening.

**C.N.:** It is planted near temples and in home gardens.

## 179. *Nymphaea pubescens* Willd.

**Family:** Nymphaeaceae

**Local Name:** Kamal

Aquatic, floating, rooted, perennial herbs; leaves orbicular or reniform, glabrous, prominently nerved beneath; flowers solitary, red, pale-rose or whitish; fruits fleshy, globose, ripening under water, seeds small, ovoid with transparent white aril.

**Sacred Aspect**

I. Flowers used in the worship of god Ganesh at Padmalaya temple.

II. Flower is used in the worship of Laxmi at Laxmipujan festival.

III. Flowers are used for worshiping Yahamogi mata (Tribal goddedd) and Mahadeva, a Hindu god.

**Miscellaneous Uses**

I. Petal, leaf petioles and the roots are edible and used as in vegetable.

**C.N.:** It is found in ponds near temples or hill stations.

## 180. *Nymphaea nouchali* Burm. f.

**Family:** Nelumbonaceae

**Local Name:** Kamal, Kamay

Perennial aquatic herbs; leaves orbicular or elliptic-sagittate, peltate, glabrous; flowers white or pale blue to white, solitary; fruits globose, seeds striate.

**Sacred Aspect**

I. Plant is observed in the garden or water bodies of several sacred places.

II. Flowers are used in the worship of Laxmimata and lord Ganesha.

**C.N.:** Various species of the two genera *viz. Nelumbo* and *Nymphaea* are called 'Kamal' and all are thoughts auspicious.

## 181. *Ocimum basilicum* L.

**Family:** Lamiaceae

**Local Name:** Sabja

Erect, much branched, aromatic herbs; leaves ovate-lanceolate; flowers white or purple; nutlets ellipsoid or oblong, black.

### Sacred Aspect

I. It is planted near the temples.

### Medicinal Uses

I. A cup of decoction of leaves is prescribed, two teaspoon thrice a day to check cough.
II. Leaf paste is topically applied on forehead to treat headache.
III. A cup of extract of leaf diluted with a cup boil water and a teaspoon honey is recommended a cup twice a day in the treatment of stomach-ache.
IV. Two teaspoon seed powder is diluted with a glass water and a pinch of common salt is adviced, a cup of it thrice a day in the treatment of dysentery.

## 182. *Ocimum gratissimum* L.

**Family:** Lamiaceae

**Local Name:** Ram Tulasi

Perennial, tall undershrubs, branches obtusely 4-angular, hispidulous; leaves ovate-lanceolate, acute, pubescent on both sides; flowers pale green or greenish yellow; nutlets subglobose, brown.

### Sacred Aspect

I. Plant is auspicious and offered in the worship of Lord Rama.
II. On the a day of Akadashi women offer 1008 inflorascences to Lord Vishnu and Lord Krishna or Lord Vitthal.
III. Inflorescence used in the worship of Marimata, a local deity.
IV. Plant is found at many sacred places.

### Medicinal Uses

I. A cup of leaf decoction. It is administered a teaspoon thrice a day, in treating stomach-ache.
II. Leaf paste is topically applied on the wounds and injuries.

**C.N.:** This species is green colour and hence are said to be Ram tulas.

## 183. *Ocimum tenuiflorum* L.

**Family:** Lamiaceae

**Local Name:** Tulasi, Krishnatulas

Erect, much branched undershrubs, stem and branches usually purplish, clothed with soft spreading hairs; leaves elliptic-oblong, obtuse or acute; flowers pale pink, nutlets slightly more than 1mm long, ellipsoid, smooth, brown.

### Sacred Aspect

I. It is the most important sacred plants of Hindus.

II. At the time of Ashadi Akadashi Lord Vitthal is decorated with a garland of inflorescence.

III. For the invitation to all gods, the plant is planted in front of the main door.

IV. To keeps evil spirit away from house, the plant is planted in front of the door.

V. Inflorescence of the plant is used in the worship of Lord Krishna, Lord Vishnu, Goddess Durgamata and Laxmi.

VI. It is believed that starting of the day by watering and worshiping tulsi plant is good sign for the family health and wealth.

### Medicinal Uses

I. Eating a fresh leaves, twice to thrice a day, is beneficial to maintain good immunity.

II. A cup of decoction of leaves is diluted with a glass boiled water. It is administrated, two teaspoon thrice a day in the treatment of cough and fever.

III. Two drops of leaf decoction are recommended in ear problems thrice a day.

IV. Leaf paste is topically applied on head in headache.

V. Leaf paste is topically applied on roughness of facial skin.

VI. Leaves are chewing as mouth freshener and against throut complaints.

## 184. *Opuntia elatior* Mill.

**Family:** Cactaceae

**Local Name:** Nivdung

Large, succulent, bushy shrubs, stem phyllodic, jointed; leaves ovate-oblong, flat, green; flowers erectopatent, at first yellowish with a red tinge; fruit fleshy, slightly depressed-truncate at apex, dark red when ripe, marked with areoles.

**Sacred Aspect**

I. It is found as impenetrable hedge around the gardens and temples.

**Medicinal Uses**

I. Stem extract is applied on burns and boils until cure.

II. Stem extract is also recommended topically on cuts and wounds until cure.

**C.N.:** It is exotic, native of South America.

## 185. *Oroxylum indicum* (L.) Vent.

**Family:** Bignoniaceae

**Local Name:** Tetu

Soft wooded trees, light brown bark; leaves bi or tri pinnate, leaflets ovate or elliptic, glabrous; flowers fleshy, dark red, compact; capsules woody, flat, brown, linear-lanceolate, seeds numerous, discoid, white, papery.

**Sacred Aspect**

I. Plant is found at many sacred places.

**Medicinal Uses**

I. A cup of decoction of the bark powder is diluted in a glass of boiled water. It is prescribed, two teaspoons daily morning and evening as a health tonic.

II. Leaf paste is topically applied on the pimple problems.

## 186. *Oryza sativa* L.

**Family:** Poaceae

**Local Name:** Tandul, Dhan

Variable, erect, stout or slender annual with varying height and form, stem tufted, hollow smooth and hairless; leaves parallel veined, ligulate, inflorescence erect or nodding loose panicle, spikelet about 7 mm long, flat.

**Sacred Aspect**

I. Sweet paste prepared from grains called 'Kheer' is served in the Buddha Pournima festival as Kheer-dan in Buddha community.

II. During the ceremony 'Dohale Jewan' a pregnant woman is offered dinner of sweets. She is wears flower garland. Every invited lady offers her pouch of rice grains and as a honour and bless to baby in a womb.

III. Seeds are used as an auspicious element in various worships.

IV. At the time of marriage rice grains are showered as Akshata on the marrying couple as a symbol of blessings and fertility.

V. Matoba is a deity related to agriculture. Rice grains and turmeric are used during worship.

VI. Nagdev idol is carved on a stone. This diety is worshiped by Kokani (Kokana). During worship, a heap of rice grains is raised before the idol. A pot of the milk and coconut is also offered.

VII. Plant of rice is worshipped by the tribals on day of Dasera festival. These plants are made into a wreath (called 'Toran') which is then worshiped outside the village or on the banks of a river.

**Medicinal Uses**

I. Cooke rice prepared is generally advised to consume during the illness.

## 187. *Oxalis corniculata* L.

**Family:** Oxalidaceae,

**Local Name:** Ambuti

Procumbent or diffuse herbs; leaves 3-foliate, glabrous, leaflets obcordate with narrow sinus, sessile; flowers are yellow, capsules linear oblong, beaked, seeds numerous, ovoid or ellipsoid, brown or reddish brown.

**Sacred Aspect**

I. It is a weedy species and naturally grows near temples.

**Medicinal Uses**

I. Decoction of leaves, two teaspoon honey and ginger (*Zingiber officinale* Rosc.) paste is recommended for the treatment of fever and cough until cure.

II. Paste of the leaf is topically applied on insect and scorpion bites.

III. Extract of whole plant is applied topically on snake bite.

IV. Leaf paste is also applied topically on rheumatic joints and swellings.

**Miscellaneous Uses**

I. Whole plant is edible and used in diet as salad it is also used as a vegetable also.

**C.N.** : It is exotic, native of Europe.

## 188. *Pandanus odoratissimus* L. f.

**Family:** Pandanaceae

**Local Name:** Kevada

Small, dioecious trees, stem supported by basal prop roots; leaves glaucous-green, ensiform, coriaceous; male flower spadix with numerous subsessile spikes, creamy white, fragrant.

**Sacred Aspect**

I. It is planted in the garden of several temples.

II. Leaves are offered to Godess Mahalaxmi.

**Miscellaneous Uses**

I. Fragrant bract of inflorascence is used for hair dress by the ladies.

**C.N.:** It is found under cultivation exclusively.

## 189. *Pedalium murex* L.

**Family:** Pedaliaceae,

**Local Name:** Mothi-Gokhru, Ubha Gokharu

Simple or branched, erect or diffuse herbs; leaves fleshy, glabrous, elliptic-oblong; flowers pale yellow, axillary; drupes pyramidal ovoid, indehiscent with 4 spines.

**Sacred Aspect**

I. It is found at many sacred places and in temple area.

**Medicinal Uses**

I. Two teaspoonful powder of fruits is diluted in two cups of water. It is boiled adding two teaspoon of honey. Two teaspoon thrice a day is advised in the treatment of diarrhoea.

II. A teaspoon fruit powder diluted in two cup boiled water a teaspoon honey and decoction of tulsi leaves(*Ocimum tenuiflorum* L.) are prescribed to check stomach ache.

III. Fruit paste topically applied on rheumatism and swellings.

IV. A teaspoon powder of fruits is mixed in a glass. Two teaspoon of. It is recommended at every night to increase sex ability and improve fertility.

**C.N.:** It is exotic, native of America. It is found on places near temples.

## 190. *Pergularia daemia* (Forssk.) Chiov.

**Family:** Asclepiadadeae

**Local Name:** Utaran, Rota

Woody twiners, stem terete, clothed with spreading white hairs; leaves reniform-ovate, cordate with rounded lobes; flowers in axillary, in peduncled corymbose cymes or paniculate cymes, dull greenish brown or white; follicles strongly reflexed, seeds broadly ovate, truncate at apex, silvery white.

**Sacred Aspect**

I. It is found on the hedges near the temples.

**Medicinal Uses**

I. A cup of decoction of whole plant is recommended for blood purification for 15-17 days.
II. A cup floral extract is diluted in a cup warm water with two teaspoon honey or sugar. It is recommended two teaspoon thrice a day, to cure cough for a week.
III. Paste of whole plant is topically applied on insect bites.
IV. Leaf paste is applied on cuts and wounds.

## 191. *Pennisetum americanum* (L.) K. Schum.

**Family:** Poaceae

**Local Name:** Bajaras, Bajari

Tall robuss annuals, culms tufted, erect, branched, noded; leaves sheathed, linear, 60-90 cm long; inflorescence cylindrical, spikelike, 15-30 cm long, dense, grain ovoid, grey, smooth.

**Sacred Aspect**

I. A bullock festival called 'Pola' is celebrated. They are worshipped on this day and are not employed for any physical work. Earthen made bullock idols are placed on the grains spread on a carpet. They are also worshipped.

**C.N.:** It is exotic, native of Western Tropical Africa. It is cultivated for food grains.

## 192. *Passiflora edulis* Sims.

**Family:** Passifloraceae

**Local Name:** Krishna Kamal

Extensive linas, tendrils simple; leaves deeply trilobed, petiolate; flowers bluish-white, solitary, axillary, corona 4-5 whorled.

**Sacred Aspect**

I. Flowers are used on the day of Janmashtami (Birthday of Lord Krishna) by Hindus.

**C.N.:** It is exotic, native of Brazil. It is cultivated in gardens as an ornamental plant.

## 193. *Phoenix dactylifera* (L.) Roxb.

**Family:** Arecaceae

**Local Name:** Khajur

Dioecious trees, trunk clothed with the persistent bases of petioles; leaves pinnatisect, greyish-green, glabrous, segments, fasciculate, rigid; flowers creamy-white in spadix, spadix 30-50 cm long, erect; drupes oblong-ellipsoid, orange-yellow.

**Sacred Aspect**

I. Dried fruit is used in various worships by in Hindus.

II. During the month of Ramzan Muslims observe the fast called Roja after complete observance and they consume dates.

III. Fruits are also used as fast food by Hindus.

**Medicinal Uses**

I. Consumption of 5-6 dried fruits every day is recommended to improve immunity, to increase haemoglobin and early recovery of good health.

**C.N.:** It is not cultivated in this region. Ripe fruits are sold in local market. It is exotic, native of Persian Gulf.

## 194. *Phyla nodiflora* (L.) Greene

**Family:** Verbenaceae

**Local Name:** Mundi

Prostrate, perennial herbs; leaves subsessile, obovate, cuneate, serrate in upper half; flowers white or pale pink; drupes globose, oblong, smooth, hard and brown.

**Sacred Aspect**

I. It is found on marshy places near temples.

**Medicinal Uses**

I. A teaspoon powder of whole plant is mixed in a cup of warm water. A cup of it twice a day is recommended in the treatment on fever and malaria for a week.

**C.N.:** It is exotic, native of South America (Stewart, 1972).

## 195. *Phyllanthus acidus* (L.) K. Skeels

**Family:** Euphorbiaceae

**Local Name:** Rai-awla

Large shrub or small tree, dense and bushy, thickish and tough main branches; leaves alternate, ovate or lanceolate, short petioles leaves thin, green, smooth on the upperside and blue-green beneath; flowers unisexuals, small, pinkish, appear in clusters; fruits numerous, oblate, densely clustered, pale yellow or white, waxy, crisp and juicy, sour. 4 to 6 seeded.

**Sacred Aspect**

I. Fruits are used in the worship of Yahamogimata, a tribal deity.
II. Fruits are also used in the worship of Durgamata in Navaratri festival.

**Medicinal Uses**

I. Root extract is topically applied scabies and herpes.
II. Leaf juices recommended at every morning to lose the weight.
III. A cup of decoction of leaves and roots is diluted with a cup of boiled water. It is administered a cup in snake bite treatment.

**C.N.:** It is exotic, native of Malaya Islands and Madagascar. It is found under cultivation exclusively.

## 196. *Phyllanthus amarus* Schumach. and Thonn.

**Family:** Euphorbiaceae

**Local Name:** Bhui Awla

Erect annuals, branches quite glabrous, spreading; leaves variable, subsessile, oval-oblong, obtuse or acute; flowers axillary; capsule smooth, seeds trigonous, longitudinally ribbed, pale brown.

**Sacred Aspect**

I. Plant is found at many sacred places.

**Medicinal Uses**

I. A cup of leaf decoction is diluted with a glass water and a pinch of common salt is added. It is recommended a cup thrice a day, in case of dysentery.
II. A cup of decoction of whole plant and a teaspoon honey is prescribed, two teaspoon thrice a day, in the treatment of fever.
III. A cup of extract of whole plant teaspoon honey and half teaspoon jeera (*Cuminum cyminum* L.) powder is recommended, two teaspoon thrice a day, for the treatment of urinary tract infection for 4-5 days.

## 197. *Physalis micrantha* Link.

**Family:** Solanaceae

**Local Name:** Koyi

Erect or prostrate, branched, glabrous herbs; leaves ovate, acute or acuminate; flowers pale yellow; berries globose, bladder like calyx, 5 angled.

**Sacred Aspect**

I. It is found in near temples.

**Medicinal Uses**

I. Leaf paste is topically applied on burn and boils.

**Miscellaneous Uses**

I. Fruits are cherished by tribals.

## 198. *Piper betle* L.

**Family:** Piperaceae

**Local Name:** Paan

Stout, climbing shrubs, nodes swollen; leaves broadly ovate, acuminate, coriaceous, glabrous; flowers in interrupted, glabrous spikes; berries red.

**Sacred Aspect**

I. Leaf is used in various worships and rites by Hindus.

**Medicinal Uses**

I. Leaf paste is topically applied on the wounds and cuts as an antiseptic.

II. Leaf chewing with arecanut is recommended for digestion and blood purification.

III. Leaf extract is used as an eye drop against eye infection.

IV. Leaf decoction diluted with a cup water. Two teaspoon honey and ginger (*Zingiber officinale* Rosc.) is prescribed, two teaspoon twice a day, in the treatment of malaria until cure.

**C.N.:** It is not cultivated in this region. Leaves are sold in market. It is exotic, native of Bali and East Indies.

## 199. *Plumbago zeylanica* L.

**Family:** Plumbaginaceae

**Local Name:** Chitrak

Erect or straggling undershrubs, stem woody, striate; leaves ovate-lanceolate, acute; flowers white or pale blue; capsules oblong, seeds oblong and smooth.

**Sacred Aspect**

I. Plant is found at many sacred regions.

**Medicinal Uses**

I. A cup of decoction of roots is diluted with a glass water, A teaspoon honey and a lime [*Citrus aurantifolia* (Christm.)Sw.] juice and a pinch of common salt is recommended in the treatment of dysentery until cure.

II. A cup of decoction of stem and leaves is prescribed in the treatment of malaria for three days.

III. Root paste is topically applied on the ringworm, rheumatism and swellings, wounds and cuts as an antiseptic.

**C.N.:** It is exotic, native of Africa (Bailey, 1949).

## 200. *Plumeria rubra* L. forma *acuminata* (Ait.) Santapaul and Irani ex Shah

**Family:** Apocynaceae

**Local Name:** Pandhara Chapha

Deciduos, small trees, bark rough, latex white; leaves whorled, oblong-lanceolate, large, prominently veined; flowers white with pale-yellow centre, fragrant, follicles paired.

**Sacred Aspect**

I. It planted at many sacred places.

II. It is cultivated near temples as ornamental, evergreen tree.

III. Leaves and flowers are used in worship of Lord Shiva.

**Medicinal Uses**

I. A teaspoon powder of bark is mixed with a cup of boiled water. It is prescribed in the treatment of malarial fever.

II. Leaf paste is topically applied on swellings, wound as well as cuts.

**C.N.:** It is found under cultivation exclusively. It is exotic, native of Tropical America.

## 201. *Polyalthia longifolia* (Sonner.) Thw.

**Family:** Annonaceae

**Local Name:** Ashok

Middle sized trees, bark light blackish brown; leaves simple, lanceolate, shining; flowers pale green; berries yellowish, seeds shining.

**Sacred Aspect**

I. It is planted at many sacred places as an ornamental tree.

II. Leaves are used in decoration of pandals mainly for religious programme.

III. Leaves are used for making the festoons and bouquet on the auspicious occasions.

**C.N.:** It is planted as an avenue tree around the temples.

## 202. *Pongamia pinnata* (L.) Pierre

**Family:** Lamiaceae

**Local Name:** Karanj

Small or middle sized trees, bark grey-white or light black; leaves ovate oblong or elliptic; flowers white or pale pink; pods obliquely oblong, compressed, woody, glabrous, flesh-brown.

**Sacred Aspect**

I. Flowers are used in worship of Gramdevta on a Bhandara occasion.

II. Plant is found in Manudevi, Unapdev and Patanadevi forest.

**Medicinal Uses**

I. Decoction of fruits diluted with two cup of boiled water, jaggery, a pinch of common salt and little ginger (*Zingiber officinale* Rosc.) paste is recommended. Three teaspoons thrice a day in the treatment of cough until cure.

II. Seed oil is topically applied on scabies, swelling and rheumatism.

III. Leaf decoction is diluted with a glass water. Leaf decoction of Pudina (*Mentha arvensis* L.) and a teaspoon honey is added in it, a cup of it twice a day in the treatment of stomach ache until cure.

IV. Seed oil two drops thrice a day, are poured in ears to check ear infection.

**C.N.:** It is wild, in the forest areas, but also planted near public places like temples, and along roadsides.

## 203. *Prosopis cineraria* (L.) Druce

**Family:** Mimosaceae

**Local Name:** Saundad, Samadi

Tall trees, branches slender, armed; leaves pinnate, leaflets oblong, small, oblique; flowers creamy-white, small, in peduncled axillary spikes; pods glarous.

**Sacred Aspect**

I. The tree is worshiped and prayed by Pawara people to stop crying of their infants.

II. A garland of fruits is offered to goddess Gaurai (Gavarai) at the festival of Aakhaji.

## 204. *Psidium guajava* L.

**Family:** Myrtaceae

**Local Name:** Peru

Small or medium sized trees, bark smooth, greenish-grey; leaves opposite, sessile, ovate-lanceolate or elliptic-oblong; flowers white; berries pyriform, variable in size, green or pale-yellowish white, embedded in white flesh, rarely flesh pink.

### Sacred Aspect

I. Fruit is used in worship of Lord Ganesha.

### Medicinal Uses

I. Consumption of fruit is recommended to increase immunity.
II. Decoction of leaves is diluted with two cups of water. Three teaspoon honey and a pinch of common salt are added. It is recommended a teaspoon thrice a day, in the treatment of fever until cure
III. Leaf decoction diluted, a lemon [*Citrus aurantifolia* (Christm.) Sw.] extract, a pinch of commen salt and three teaspoon honey is recommended in the treatment of dysentery and diarrhoea.

**C.N.:** It is exotic, native of Tropical America. It is found under cultivation only.

## 205. *Pterocarpus marsupium* Roxb.

**Family:** Fabaceae

**Local Name:** Bibla, Bijja

Moderate sized trees, bark yellowish grey or ash coloured; leaves 5-foliate, ovate-oblong or ovate-lanceolate; flowers yellow, yellow pubescent racemes; pods nearly circular, 1 seeded, flat.

### Sacred Aspect

I. Wood of the plant is used in 'Havan' purpose.
II. Fruit is tied to small babies and healthy domestic animals to protect them from evil eye.
III. Tree is found at Manudevi, Kondaibari, Toranmal and Pal forest.

### Medicinal Uses

I. Wood powder is diluted in a glass boiled water and two-three teaspoon honey is recommended orally in a joint pain for a week.
II. Leaf paste is topically applied on the boils as a coolant and antiseptic.
III. Leaf paste is diluted with a pinch of turmeric (*Curcuma longa* L.) powder and coconut (*Cocos nucifera* L.) oil is topically applied on pimples.

IV. Leaf paste is topically applied on cuts and wounds.

V. Wood powder is mixed in a glass water. It is recommended in the digestion problems for a week.

## 206. *Pueraria tuberosa* (Roxb. ex Willd.) DC.

**Family:** Fabaceae

**Local Name:** Bhui Kohala

Extensive twiners with large tuberous roots; leaves 3-foliate, ovate, acuminate, entire; flowers bluish-purple; pods 1-6 jointed, flat, brownish tomentose, seeds reddish brown, elliptic-oblong and smooth.

**Sacred Aspect**

I. Plant is found at many sacred places.

**Medicinal Uses**

I. Tuber paste is topically applied on rheumatism and swellings.

II. Seed powder paste diluted with coconut oil is applied on cuts and wounds also.

**Miscellaneous Uses**

I. Powder obtained from dried tubers is added in bread and consumed. It is believed to be tonic for better health.

## 207. *Punica granatum* L.

**Family:** Punicaceae

**Local Name:** Dalimb

Deciduous shrubs or small trees, bark dark grey; leaves opposite, ovate or elliptic oblong; flowers orange-red, showy; berries orange sized, numerous seeds, angled, aril filled with acidic or sweet juice.

**Sacred Aspect**

I. Fruit is used in worship of Lord Ganesha and Satyanarayana.

**Medicinal Uses**

I. Consumption of fruit is recommended to increase white blood cells.

II. Leaf paste is applied on boils as a coolant.

III. Dry seed powder is mixed with milk cream. It is topically applied on pimples.

**Miscellaneous Uses**

I. Fruit is edible and one of the most important fruits in India.

**C.N.:** It is exotic, native of Afghanistan and Persia.It is found only under cultivation.

## 208. *Quisqualis indica* L.

**Family:** Combretaceae

**Local Name:** Madhumalti

Large straggling much branched linas; leaves elliptic or elliptic-oblong, acuminate; flowers showy, fragrant, whitish or deep red or pink, sessile corymbose spikes; fruits mostly not developed.

**Sacred Aspect**

I. It is planted in temple gardens.

**Medicinal Uses**

I. Leaf paste is topically applied on insect bites.
II. Juice of flowers is daily, a glass at night is given to control blood sugar.
III. Leaf decoction diluted with a glass boiled water and two teaspoon honey is recommended in the treatment of malarial fever till cure.

**C.N.:** It is exotic, native of tropical Asia. It is found only under cultivation for fragrant, attractive flowers.

## 209. *Raphanus sativus* L.

**Family:** Brassicaceae

**Local Name:** Mula

Annual or biannual with fusiform tuberous roots; leaves radical, membranous; flowers white or lilac, dark veined, siliquae long and spongy, seeds globose, reticulately veined, single seriate.

**Sacred Aspect**

I. Vegetable prepared from the leaves is used in food offering to goddess Mahalaxmi in the Mahalaxmi worship.

**Medicinal Uses**

I. Radish is recommended in daily diet for treating constipation.

**Miscellaneous Use**

I. Leaves and pods are cooked as vegetable.

**C.N.:** It is exotic, native of Europe.

## 210. *Ricinus communis* L.

**Family:** Euphorbiaceae

**Local Name:** Erand

Large, erect shrubs or small trees, stem glaucous; leaves large, peltate, orbicular; flowers monoecious, in terminal panicles, staminate flowers below and pistillate above; capsules depressed, globose, seeds oblong-ovoid, smooth, shining, white or creamy.

**Sacred Aspect**

I. Plant is used in Holi festival as the symbol of Holika. In tribal community, Holi is celebrated for many days. In this festival a branch of the tree is placed at the middle of the common place. Fuel wood is arranged around it. The head person of the tribal community worships and ignites it and all his followers dancing around it.

**Medicinal Uses**

I. Seed oil is topically applied on the forehead against headache.

II. Leaf paste diluted with coconut (*Cocos nucifera* L.) oil and a pinch of cumin seeds (*Cuminum cyminum* L.) applied in case of ringworm.

III. Leaf paste diluted with groundnut (*Arachis hypogaea* L.) oil is tropically applied on rheumatism.

**C.N.:** It is exotic, native of tropical Africa. It is found under cultivation and also naturalised as an escape from cultivation.

## 211. *Rosa damascena* Mill.

**Family:** Rosaceae

**Local Name:** Gulab

Scandent shrubsbearing stout hooked prickles; leaflets usually 5, ovate, oblong, smooth above; flowers singled or double on an upright pedicel, fragrant, red to dark red, hypanthium glandular-hispid.

**Sacred Aspect**

I. Flowers are used in worship of Bhagwan Mahaveer.

II. In the Muslim Dargah and Majar the flowers are used for the making 'Chadar' offering to their god.

III. Flowers are used for welcome of guests and honourable eminents.

**Miscellaneous Uses**

I. Decoction of flowers mixed with Multani soil and used as face pack topically on the face.

**C.N.:** It is exotic, native of East Asia. It is found only under cultivation.

## 212. *Saccharum officinarum* L.

**Family:** Poaceae

**Local Name:** Uss

Perennial, clumps consisting of a number of strong unbranched stems, nodes present, the alternate leaves, leaves green, thick midribs, saw-toothed edges, terminal inflorescence, panicles, spikelets borne on side branches, silky hair, fruits dry,

**Sacred Aspect**

I. Cane is used for making the small tent meant for the worship of Khandoba at the ceremony of Gondhal Jagaran, before marriage ceremony.

II. Cane is used for making the tent of the marriage of Tulsi Vivah after Diwali. The marriages are decided after this day in Hindus.

III. Cane is used to decorate 'Tagatrao' (procession of gods), a vehicle meant for religious purpose.

IV. Dongardev is a village deity of tribals *viz.* Bhil and Kokani. Worshiper raises a tent in common place using culms of sugarcane, branches of *Tagetes erecta* L. and *Ocimum tenuiflorum* L. A heap of rice grains is raised around this tent. During worship dates, grains of finger millet (Nagali) and coconut are used.

**Medicinal Uses**

I. Sugarcane juice is popularly used as a cold and neutritious drink.

## 213. *Salvadora persica* L.

**Family:** Salvadoraceae

**Local Name:** Pilu

Evergreen trees, trunk straight or crooked, bark dull-grey or greyish white or greyish brown; leaves glabrous, acute, fleshy; flowers pink to deep purple; fruits 1-seeded, pinkish.

**Sacred Aspect**

I. Ripe fruits are considered as Ranmeva in worship of goddesses Marimata.

**Medicinal Uses**

I. A glass of decoction of leaves is diluted with 100 ml water. It is prescribed a teaspoon thrice a day, in the treatment of cough for the 4-5 days.

II. Twigs are used as toothbrush in tooth problems and foul smell of mouth.

III. Seed oil is used for the massage in rheumatism treatment.

## 214. *Santalum album* L.

**Family:** Santalaceae

**Local Name:** Chandan

Small evergreen trees, drooping branches, bark blackish; leaves elliptic-lanceolate, subacute, glabrous, shining green above; flowers brownish purple; drupes subglobose, purplish black when mature.

**Sacred Aspect**

I. Wood is used during Havan (sacred fire) by Hindus.
II. Wood is used in the funeral rites in Hindu religion.
III. Wood paste is applied as small spot called Tikka on forehead of Lord Vitthal and Lord Vishnu and Lord Mahadev.
IV. Wood paste is applied on idols of gods.

**Medicinal Uses**

I. Bark paste is topically applied on the face for skin fairness and softness.
II. Bark paste is applied on the body parts as coolant.

## 215. *Sapindus emarginatus* Vahl

**Family:** Sapindaceae

**Local Name:** Ritha

Trees, bark brown, longitudinally fissured; leaves compound, leaflets ovate-oblong; flowers small, white in terminal panicles; fruits mostly 3-loabed.

**Sacred Aspect**

I. Plant is found at sacred places like Toranmal, Patanadevi, Pal *etc.*

**Medicinal Uses**

I. Bark paste is topically applied against body ache and headache.
II. Bark paste diluted with coconut (*Cocos nucifera* L.) oil is topically applied to treat ringwarm.

**Miscellaneous Uses**

I. Fruit pulp is used as detergent.

## 216. *Saraca asoca* (Roxb.) de Wilde

**Family:** Caesalpiniaceae

**Local Name:** Ashok

Evergreen trees, bark dark grey or light black; leaves pinnate, rough, leaflets 6- pairs, oblong; flowers scarlet; pods subtorulose, pulpy, seeds 3-12, brown, compressed, glabrous.

**Sacred Aspect**

I. Wreaths of leaves are used for decoration on auspicious events and festivals.

II. Leaves are used for making the bouquets and garlands for offering to gods.

**Medicinal Uses**

I. About a cup decoction of flowers and leaves diluted with two cups of water is prescribed as a health tonic.

C.N.: It is not denizen of Khandesh region. It is found under cultivation only.

## 217. *Schleichera oleosa* (Lour.) Oken

**Family:** Sapindaceae

**Local Name:** Kusum

Deciduous trees; leaves abruptly pinnate, leaflets in 2-4 pairs, elliptic-lanceolate; flowers creamy white or pale yellow; fruits ovoid, pointed, dark brown, seeds oblong, smooth and glabrous.

**Sacred Aspect**

I. It is found near local deities at Dhadgaon and Toranmal.

II. Young leaves are used for making the festoons and for decoration of pandals on auspicious social event.

**Medicinal Uses**

I. Seed oil is topically applied on the burns and cuts as a coolant.

II. Fruit pulp is used for the nourishment and shining hair topically.

III. Paste of bark powder is applied topically on itching.

IV. Seed oil is applying topically on ringworm.

**Miscellaneous Uses**

I. Seed oil is used for the polishing the furniture and photoframes.

## 218. *Securinega virosa* (Roxb. ex Willd.) Baill.

**Family:** Euphorbiaceae

**Local Name:** Pandharphali

Bushy, straggling shrubs, bark light brown; leaves obovate, elliptic-oblong; flowers minute, greenish to creamy yellow; fruits globose, white when ripen, seeds rounded on the back, minutely puncate.

**Sacred Aspect**

I. Plant is found at many sacred places.
II. Fruits are consumed as Ranmeva during tribal festivals.

**Medicinal Uses**

I. Fruit pulp about half spoon is advised twice a day while treating of stomach ache.
II. Leaf paste is applied topically to treat the rheumatism and swellings.
III. The fractured portion of bones is tightly covered by leaves and cotton cloth for a day at every morning for a month.

**C.N.:** It is grown as a hedge around temples.

## 219. *Semecarpus anacardium* L.f.

**Family:** Anacardiaceae

**Local Name:** Bhilava

Tall, deciduous trees, bark brown black; leaves coriaceous, obovate, oblong, rounded at apex; flowers small, born panicles 15-30 cm long, terminal, grey tomentose, nuts reniform, dark-brown or nearly black.

**Sacred Aspect**

I. Plant is found at many sacred places.
II. Fruit is used in black magic.

**Medicinal Uses**

I. Fruit extract about a cup is recommended thrice a day to check asthma for 12-15 days.
II. Seed oil is applied on swelling and rheumatic joints.
III. Seed oil is applied on injuries caused due to thorns.

**C.N.:** It is found under cultivation exclusively.

## 220. *Sesamum orientale* L.

**Family:** Pedaliaceae

**Local Name:** Til

Glandular pubescent, erect herbs; leaves broadly linear-lanceolate or elliptic-lanceolate; flowers pink, purple or white; capsules oblong or ovoid, seeds mostly white or rarely brown or black.

**Sacred Aspect**

I. Seeds are most important part of Makar Sankrati festival. The dried seeds are mixed with jaggery or sugar and served to relatives with the slogan 'Tilgul Ghya God God Bola'.

**Medicinal Uses**

I. Seed oil is topically applied on cuts, wounds and boils and also used for massage.

**Miscellaneous Uses**

I. Seeds are edible and used in the preparation of various sweetmeats and culinary preparation.

II. Seed oil is used for culinary purpose.

**C.N.:** It is a cultivated crop as source of edible oil. It is exotic, native of Africa (Dogra, 2011).

## 221. *Sesbania grandiflora* (L.) Poir.

**Family:** Fabaceae

**Local Name:** Hadga

Soft wooded trees; leaves pinnate, leaflets 10-30 pairs, oblong-obovate; flowers large, showy, white or cream-coloured, tinged with pink, papilionaceous; pods reaching 30 cm length, woody pendent, seeds 17-20, oblong-ellipsoid, brown, compressed, smooth.

**Sacred Aspect**

I. It is found near the temple area.

**Miscellaneous Uses**

I. Flowers are edible and used as vegetable.

**C.N.:** It is exotic, native of Indonesia. It is found under cultivation only.

## 222. *Sesbania sesban* (L.) Merr.

**Family:** Fabaceae

**Local Name:** Sheuri, Shewari

Medium sized trees; leaves pinnate, leaflets oblong; flowers yellow, born in racemes, papilionaceous; capsules elongated, seeds many, glossy, brown.

**Sacred Aspect**

I. Plant is found at many sacred places.

**Medicinal Uses**

I. Leaf and flower decoction about a glass is mixed with four teaspoon of honey. It is recommended for a month twice a day for improving fertility.

**C.N.:** It is found only under cultivation especially as a wind breaker around crop fields, farmyards *etc.* It is exotic, native of South America (Martin *et al.*, 1987).

## 223. *Setaria italica* (L.) P. Beauv.

**Family :** Poaceae

**Local Name :** Bhandi

Annual herb, culms erect, tall, noded; leaves lanceolate, flat, sheathed, ligulate; panicles spiciform, dense and interrupted; grains ovate, grey-brown.

**Sacred Aspect**

I. Indal festival is celebrated by the tribals. Thereby, they worship the god Indra. A day is decided by the people of village. The tribals apply turmeric powder (*Curcuma longa* L.) to the grains of Bhandi and distribute to the people and thereby invite them for Indal festival.

**Miscellaneous Uses**

I. Straw is used as fodder for cattle.

II. Straw is also used in making partition walls of abodes by the tribals.

**C.N.:** It is exotic, native of Near East (China) (Singh and Nigam, 2017).

## 224. *Sida acuta* Burm. f.

**Family:** Malvaceae

**Local Name:** Bokya

Much branched, undershrubs; leaves oblong lanceolate or elliptic, sessile; flowers yellow or pale orange; schizocarps not exceeding calyx, mericarp pale brown, seeds tri gonous, dark brown or blackish-brown, smooth, glabrous.

**Sacred Aspect**

I. Plant is found at many sacred places.

**Medicinal Uses**

I. Fruit pulp is applied on the burns and boils as a coolant.

II. Root decoction diluted with three cup of boiled water and two teaspoon honey is recommended for the treatment of stomach-ache for four days. It is also benefits to increase immunity.

III. Decoction of leaves about a cup is diluted with a glass water and a teaspoon lime juice added. It is advised at every night on gastric problems for a month.

**C.N. :** It is exotic, native of Tropical America.

## 225. *Sida cordata* Burm.f.

**Family:** Malvaceae

**Local Name:** Bala

Erect herbs; leaves simple, serrate or crenulated-dentate; flowers yellow, fasciculate or solitary, axillary; mericarps 5, thin walled, seeds glabrous, brown.

**Sacred Aspect**

I. Plant is found at many sacred places.

**Medicinal Uses**

I. Paste of entire plant is topically applied on rheumatic joints and swelling until cure.
II. Root paste is applied on the cuts and wounds.
III. Root decoction about two teaspoon with a teaspoon honey is prescribed thrice a day, to treat stomach ache.

**C.N.:** It is exotic, native of South America.

## 226. *Sida spinosa* L.

**Family:** Malvaceae

**Local Name:** Gangeti, Kantalobata

Erect, woody herbs, stellately hairy or grey tomentose in younger part; leaves ovate-oblong or ovate elliptic; flowers white, rarely yellowish white; mericarps 5, trigonous, stellate, seeds ovoid, dark brown, smooth.

**Sacred Aspect**

I. Plant is found at many sacred places.

**Medicinal Uses**

I. Root paste is topically applied on boils as a coolant.
II. Root paste mixed with coconut oil and 'Multani' soil is topically applied to treat pimples.
III. Leaf paste is used topically on swellings.
IV. Root decoction and ginger paste (*Zingiber officinale* Rosc.) is recommended to check fever. About three spoonful of it is administered, thrice a day, for four days.

**C.N.:** It is exotic, native of Pantropical region (Singh and Srivastava, 2000).

## 227. *Solanum anguivi* Lamk.

**Family:** Solanaceae

**Local Name:** Mothi ringni

Stout, erect, armed, stellately hairy undershrubs; leaves broadly elliptic or elliptic-oblong; flowers purple, drooping, berries globose, orange when ripe, seeds spherical, flat microscopically pitted.

**Sacred Aspect**

I. Fruits are offered to goddess Mahalaxmi.

**Medicinal Uses**

I. Root extract is is poured in ears. Two drops thrice a day are prescribed.

**Miscellaneous Uses**

I. Fruits are used as vegetable, by the tribals.

## 228. *Solanum melongena* L.

**Family:** Solanaceae

**Local Name:** Vangi, Vange

Armed undershrubs, stem and branches stellately grey tomentose; leaves ovate-oblong; flowers purple, in racemose cymes; berries obovoid or slender, purple, dark-purple, pure white or pure green, with large calyx.

**Sacred Aspect**

I. Vegetable fruit is used as food offering to goddess Mahalaxmi.

II. Its consumption is avoided during Chaturmas (four months in rainy season)

III. Preparation of fruits is called 'Bharit' is popularly served in many festivals, functions and ceremonies.

**Miscellaneous Uses**

I. Fruits are classically used as vegetable.

**C.N.:** It is exotic, native of East Indies (Singh *et al.*, 2001). It is cultivated for fruits in this region.

## 229. *Solanum nigrum* L.

**Family:** Solanaceae

**Local Name:** Kangani

Erect annuals; leaves ovate, ovate-oblong or ovate lanceolate; flowers small,white; berries purplish black, red or yellow, globose, smooth, shining, seeds many, yellow minutely pitted.

**Sacred Aspect**

I. Plant is found at many sacred places.

**Medicinal Uses**

I. Paste of leaves is applied on burns and boils to avoid infection.

II. Fruit paste is applied on swollen organ and rheumatic joints overnight.

III. Leaf decoction about a cup mixed with three teaspoon lime juice and three teaspoon honey prescribed against dysentery.

IV. Extract of whole plant about a cup is recommended twice a day to cure jaundice. It is advised till cure.

**C.N.:** It is exotic, native of Tropical America (Debnath And Debnath, 2017).

## 230. *Solanum virginianum* L.

**Family:** Solanaceae

**Local Name:** Bhuiringani

Diffuse or prostrate annuals, zig-zag stem, clothed with stellate hairs; leaves ovate-elliptic or elliptic-oblong; flowers violet purple, berries white, with green veins, yellow when ripe, globose, seeds numerous, glabrous, ovoid-reniform, pale brown.

### Sacred Aspect

I. Plant is found at Padmalaya, Unapdev, Mehun *etc.*

### Medicinal Uses

I. Fruit extract about a cup mixed with two teaspoon sugar is given orally. Two teaspoon of it thrice a day in the treatment of paralysis.

II. Fruit paste is also massaged on the paralised body part.

III. Root paste is applied against insect bite.

IV. Extract of whole plant is also applicable on swellings.

V. Root decoction diluted with a cup of boiled water and two teaspoon honey is prescribed a teaspoon thrice a day in the treatment of fever is advised.

**C.N.:** It is exotic, native of Paleotropical region (Singh and Srivastava, 2000).

## 231. *Sorghum vulgare* Pers.

**Family:** Poaceae

**Local Name:** Jowar

Erect herbs, stem cylindrical, nodded, height from 2-15 ft, solid, central point pithy, side branches shorter than main stem axis; leaves alternating, simple, sheathed, linear lanceolate, long, flat; spiklets found in several pairs, glume enclosing two florets, lower ones sterile and upper ones perfect.

### Sacred Aspect

I. Grains are coloured with turmeric (*Curcuma longa* L.) powder and tied in yellow cloth. This cloth is tied to earthen pot ful of water. It is worshiped during marriage for prosperity and better life of marriage couple.

II. Roti (Bread) made from the seed flour is used for food offering to Gods.

III. Flour of grains is mixed with turmeric (*Curcuma longa* L.) powder. This powder is called as 'Chiksa'. It is applied to bride and bridegroom while departing from place of marriage.

IV. Roti is commonly served in the auspicious ceremony to the participants as important ingredient of lunch or dinner.

V. Roti is used for food offering to goddess Mahalaxmi in rural area of Khandesh.

VI. Ghata festival is celebrated by farmers. The food grains example of Jawar after harvesting, are used in a local food preparation called 'ghata'. It is consumed and also offered to neighbours. The first harvest of grains is thus honoured.

VII. Tribals worship Kansari goddess. After harvesting, the cobs of jawar are brought on farmyard. Grains are removed and cleaned. The heap of grains is the worshipped. After this worship, the tribals consume bread prepared from it.

**C.N. :** It is exotic, native of Africa.

## 232. *Spilanthus calva* DC.

**Family:** Asteraceae

**Local Name:** Akkalkadha

Annual, erect or ascending, 20-45 cm tall; leaves ovate-lanceolate, 3 nerved; flowering heads ovoid, yellow; achenes obovoid, black, ciliate on margins, pappus absent.

### Sacred Aspect

I. Flower buds are used in worship.

II. Plant is found at Manudevi, Unapdev and some of Toranmal Sacred groves.

### Medicinal Uses

I. Decoction of floral buds about a cup is diluted with a cup of water and a teaspoon mixed sugar in it. It is recommended two teaspoon thrice a day to the child for improving the pronunciation ability.

II. Root extract about a cup is diluted with two cup boiled water and three teaspoon honey is prescribed a teaspoon thrice a day in the treatment of cough and fever until cure.

III. Leaf paste is topically applied on ringworm, cuts and wounds as an antiseptic.

## 233. *Spondias pinnata* (L. f.) Kurz.

**Family:** Anacardiaceae

**Local Name:** Ambada

Trees, bark smooth, ash-coloured; leaves imparipinnate, leaflets opposite, subsessile, oblong-elliptic or oblong, acuminate, entire, oblique; flowers polygamous, pinkish green, drupes yellow, fleshy.

**Sacred Aspect**

I. Plant is planted at sacred places.

**Medicinal Uses**

I. Bark powder about a tablespoon diluted with a cup of warm water and pinch of sugar is added. It is recommended cup thrice a day in the treatment of stomach-ache until cure.

II. Two drops of leaf extract are poured in ear against ear-ache thrice a day until cure.

**Miscellaneous Uses**

I. Fruits are edible and pickled.

**C.N.:** It is exotic, native of Tropical Asia (Martin *et al.*, 1987).

## 234. *Sterculia urens* Roxb.

**Family:** Sterculiaceae

**Local Name:** Kadai, Kadoni, Kevda, Kudal

Medium sized tree, bark pale grey or brown; leaves large, palmately lobed, acuminate, cordate; flowers small, yellow to olive yellow, follicles 5, 3-6 cm, ovoid-oblong, seeds oblong, black, smooth.

**Sacred Aspect**

I. The plant is found at several sacred places.

**Medicinal Uses**

I. A teaspoon gum powder is diluted in a cup of water and a pinch of sugar is homogenised with it. It is advised to control dysentery and to treat joint pain.

II. Half teaspoon gum diluted with a cup of cow milk is recommended as a health tonic for a month.

**Miscellaneous Uses**

I. Seeds are edible.

## 235. *Syzygium cumini* (L.) Skeels.

**Family:** Myrtaceae

**Local Name:** Jambhul

Trees, bark ash-white coloured, branches terete, glabrous; leaves opposite, decussate, elliptic, ovate-lanceolate or oblong-lanceolate; flowers fragrant, pale greenish white; berries oblong or subglobose, dark-purple, seeds globose or oblong-cylindric.

**Sacred Aspect**

I. It is planted mostly near the temples.

II. Twigs are used for tent making during the marriage ceremony.

III. Twigs are also tied to the houses. It is called as Mandap. This is supposed to be an invitation for close of the mariage.

IV. Fruits are important part of Ranmeva which is consumed in fasting period.

**Medicinal Uses**

I. Leaf decoction is rubbed into the mouth for the treatment on blisters in mouth.

II. Two teaspoon seed powder diluted with a cup of boiled water is recommended a cup at morning and evening before the meal everyday is advised for controling the blood sugar level.

III. Seed powder two tablespoon diluted with 'Multani' soil and coconut oil is tropically applied on pimples.

IV. Fruit is edible and continued eating of fruits improves immunity.

## 236. *Tabernaemontana divaricata* (L.) R. Br.

**Family:** Apocynaceae

**Local Name:** Tagar, Chandani

Glabrous, much branched shrubs; leaves obovate-lanceolate, oblong or elliptic, pale green, shining above; flowers white.

**Sacred Aspect**

I. Plant is mostly planted near the temples.

II. Flowers are used in the worship of Lord Mahadev.

III. Flowers also used in the worship of God Muktaimata at Mehun and Muktainagar.

IV. Flowers are used for the decoration and for making the floral ornaments for worship.

**Medicinal Uses**

I. Leaf paste is topically applied on boils and burns.

II. Extract of leaf about half cup diluted with glass of water and three teaspoon honey and ginger (*Zingiber officinale* Rosc.) extract is prescribed. Two teaspoon thrice in a day in the treatment of fever and cold up to 4-5 days, is advised.

III. Leaf paste is diluted with coconut oil (*Cocos nucifera* L.) and applied topically on cuts and wounds.

## 237. *Tagetes erecta* L.

**Family:** Asteraceae

**Local Name:** Zendu

Stout, branched, erect herbs, branches ribbed; leaves aromatic; flowering heads orange to yellow; achenes linear-oblong, pappus 1-seriate, of 5-8 lanceolate scales.

**Sacred Aspect**

I. Flowers are thought auspicious their festoons are used in many festivals and decoration.

II. Floral garlands are offered to gods and goddess.

III. Flowers are used to decorate vehicles during Diwali and Dasera festival.

IV. Flowers are used in various worships.

**Medicinal Uses**

I. Leaf paste is topically applied on injuries and wounds.

II. Leaf paste is mixed with turmeric (*Curcuma longa* L.) powder and milk cream is applied on the face for fairness.

**C.N. :** It is exotic, native of Mexico. It is planted for ornamental religious purpose.

## 238. *Tamarindus indica* Linn.

**Family:** Caesalpiniaceae

**Local Name:** Imli

Large trees, bark dark-grey or light black; leaves pinnate, leaflets 6-15 pairs, oblong; flowers yellowish with pink strips; pods subtorulose, slightly falcate, pulpy, endocarp septate, mesocarp pulpy, seeds 3-12, brown, compressed, smooth, glabrous.

**Sacred Aspect**

I. Dried fruits are used for food offering to goddess Marimaata. The goddess is most popular in rural area of Khandesh.

**Medicinal Uses**

I. Seed powder about a tablespoon is diluted with coconut oil and applied topically on burns and boils.

II. Two cups of leaf decoction diluted with a teaspoon honey and a pinch of common salt is recommended two teaspoon thrice a day in the treatment of fever and cold.

III. Leaf paste applied topically on swellings.

IV. Fruit juice prepared from the fruits is preferred in sunstrokes.

V. Seed powder diluted with water and the paste is applied on the pimples.

**C.N. :** It is exotic, native of tropical Africa.

## 239. *Tectona grandis* L.f.

**Family:** Verbenaceae

**Local Name:** Sag

Large deciduous trees, bark grey-white; leaves large, elliptic or obovate, acute; flowers small, white, born in large panicles; drupes subglobose, somewhat 4-lobed, seeds oblong, brown.

**Sacred Aspect**

I. Leaves are used to cover the roof of a house to avoid the entering of evil.

II. Wood is used for making the temples or temple doors and designer furniture for the temple.

III. Wood is used for making the chariots for gods and also used for making the musical instruments.

IV. Vagdev festival is celebrated by the tribals. A branch of this tree is placed erect in a place. If a tiger looks at it, then it is believed that the tiger willnot disturb anybody.

**Medicinal Uses**

I. Bark powder about two tablespoon diluted with two cups boiled water and two teaspoon honey is prescribed to check stomach-ache.

II. Paste of wood powder is topically applied on the rheumatism and swellings.

III. Seed oil is topically used as a hair tonic.

## 240. *Tephrosia purpurea* (L.)Pers.

**Family:** Fabaceae

**Local Name:** Sarpankha

Branched undershrubs; leaves pinnate, elliptic-oblong or oblanceolate; flowers reddish purple or bright pink; pods flat, slightly curved or straight, 3-9 seeded, seeds black brown, oblong, compressed, smooth, glabrous.

### Sacred Aspect

I. Plant is found at many sacred places.

### Medicinal Uses

I. Three cups of root decoction is mixed with two teaspoon honey. Two teaspoon of it thrice a day for the treatment of cold and fever is advised.

II. Dried root powder is used as toothpaste to get relief from toothache.

III. Root paste is applied on snake and scorpion bite.

IV. Decoction of whole plant is diluted with a glass of water. Four teaspoon honey and a pinch of sugar added. Three teaspoon at every night is prescribed as a health tonic a month.

**C.N.:** It is exotic, native of Pantropical region (Singh and Srivastava, 2000).

## 241. *Terminalia arjuna* (Roxb.) Wight and Arn.

**Family:** Combretaceae

**Local Name:** Arjun

Large, much branched trees with gray bark; leaves alternate or sub-opposite, oblong, elliptic or obovate; flowers small, creamy white, fragrant; drupes ovoid or ovoid-oblong, woody, dark brown, seeds linear, glabrous.

### Sacred Aspect

I. Bhagats advise the patients for to rest under the shade of the tree. It is believed that the angels rested on the branches of the tree.

II. Plant is found at many sacred places.

### Medicinal Uses

I. A teaspoon bark powder is diluted with a cup of water. It is advised twice a day in the treatment of bone fracture up to a month.

II. Leaf paste is topically applied on wounds and cuts.

III. Root powder is used as toothpaste to treat toothache.

## 242. *Terminalia bellirica* (Gaertn.) Roxb.

**Family:** Combretaceae

**Local Name:** Beheda

Large trees, bark dark grey; leaves clustered at end of branches, broadly elliptic or elliptic obovate; flowers small, pale greenish-yellow or creamy; drupes ovoid or ellipsoid.

### Sacred Aspect

I. Dried fruit is used in the worship of Varundevata for good monsoon.

II. Plant is found at many sacred places.

### Medicinal Uses

I. Dried fruit powder about two tablespoon is diluted in two cups of boiled water. Half cup of it is recommended thrice a day to check stomach ache until cure.

II. Fruit powder about two tablespoon diluted with the two cups water, ginger (*Zingiber officinale* Rosc.) extract and three teaspoon honey is prescribed in the treatment of dry cough for a weak.

III. Fruits and seeds are consumed as a health tonic.

## 243. *Thespesia populnea* (L.) Soland. ex Corr.

**Family:** Malvaceae

**Local Name:** Paras-pimpal

Evergreen trees; leaves entire, cordate, orbicular, acuminate; flowers large, yellow, axillary, solitary; capsules globose, depressed atop, seeds ovoid, black, pubescent.

### Sacred Aspect

I. It is planted ate some sacred places.

### Medicinal Uses

I. Paste of inflorescence is topically applied on the swellings.

II. Fruit paste is diluted with the coconut (*Cocos nucifera* L.) oil and applied on scabies.

III. Leaf paste is topically applied on cuts and wounds until cure.

## 244. *Thuja plicata* ex D.Don

**Family:** Cupressaceae

**Local Name:** Vidya

Large tree; leaves scale-like, in opposite pairs, green above, aromatic; cones slender, green, yellow-green, brown when ripe, seeds with paper wings.

**Sacred Aspect**

I. Small part of the twig is placed in student's notebook as the symbol of Vidya (meaning knowledge).
II. It is planted at sacred places in gardens.
III. Twigs are used for making bouquet.
IV. Twigs are also used for the decoration on many auspicious ceremonies and festivals.

**C.N.:** It is a gymnospermic tree, planted in gardens for as avenue tree in gardens.

## 245. *Tinospora cordifolia* Edgew and Hk.f.

**Family:** Menispermaceae

**Local Name:** Gudvel

Deciduous, woody twiners with striate or lenticulate green branches, bark is loosly grey white; leaves cordate, orbicular, membranous, glabrous; flowers yellowish-green, in axillary racemes; drupes red, seeds ventrally grooved.

**Sacred Aspect**

I. Plant is found at many sacred places.

**Medicinal Uses**

I. Root decoction about a cup is diluted with a cup of warm water, a pinch of salt and two drops of lemon are mixed in it. A cup of it thrice a day is prescribed to treat loose motion.
II. Dried powder of whole plant is diluted in a glass of water. Three to four teaspoon honey is added in it. This syrup about tablespoon is prescribed twice a day for fractures.
III. Stem extract about a cup is diluted in a cup boiled water. Two teaspoon of it thrice a day, is advised in the treatment of cough until cure.
IV. Decoction of whole plant is prepared. Three teaspoon honey is added in it. Three teaspoon it is given at morning and evening regularly to increase immunity.

## 246. *Tribulus terrestris* L.

**Family:** Zygophyllaceae

**Local Name:** Gokhru, Gokshura mool

Prostrate or procumbent, pilose herbs; leaves pinnate, opposite, rarely alternate, leaflets 3-7 pairs, ovate-elliptic; flowers yellow, solitary; fruits usually 5-angled, mericarps 5, rarely 3 to 4.

**Sacred Aspect**

I. Herb is commonly found at sacred places.

**Medicinal Uses**

I. A teaspoon fruit powder is diluted in a cup of water two-three drops of ginger and honey is homogenised with it. Two spoonful of it is administrered against cough and cold.

II. A teaspoon dried fruit powder diluted in a glass of warm water. Two teaspoon of honey is added in it. It is advised to treat urinary problems.

**C.N. :** It is exotic, native of Tropical America.

## 247. *Tricholepis indica* (Willd.) Almeida

**Family:** Asteraceae

**Local Name:** Dahan

Erect herbs, stem rigid, glabrous; leaves sessile, lanceolate, linear-lanceolate, acute, margin spiny-toothed; flowering heads purple-pink, bristly, ovoid, terminal; achenes ribbed, oblong, pappus barbellate.

**Sacred Aspect**

I. Herb is commonly found at sacred places.

**Medicinal Uses**

I. Paste of whole plant is topically applied on cuts and wounds.

## 248. *Trichosanthes cucumerina* L.

**Family:** Cucurbitaceae

**Local Name:** Jangli padval

Monoecious, herbaceous, slender, tendrillar climbers; leaves orbicular reniform or broadly ovate; flowers white, males in axillary racemes while female in solitary; berries glabrous, ovoid, conical, 8-10 seeded, seeds ovate-oblong, compressed, base attenuate.

**Sacred Aspect**

I. Climber is found at many sacred places.

**Medicinal Uses**

I. Root extract about a cup, and a cup of water and two teaspoon honey is recommended thrice a day in the treatment on stomach ache.

## 249. *Trichosanthes tricuspidata* Lour.

**Family:** Cucurbitaceae

**Local Name:** Mukul

Woody, extensive climbers; leaves entire or palmately 3-5 lobed, denticulate, glabrous; flowers white; berries bright red with orange stripes when ripe, seeds ovate-oblong, smooth, emarginate, white.

**Sacred Aspect**

I. Climber is found at many sacred places.

**Medicinal Uses**

I. A teaspoon dried root powder is diluted in a cup of warm water. Two teaspoon honey is homogenised with it. It is advised in the treatment of fever for four to five days.

II. Root powder is diluted in a cup water and recommends a cup thrice a day to stop the vomits.

III. Root extract is topically applied on scorpion bite, rheumatism and swelling.

## 250. *Tricticum aestivum* L.

**Family:** Poaceae

**Local Name:** Gahu

Annual grass, culms 60-100 cm long, culm-internodes thin-walled, leaf-sheaths glabrous on surface, leaf-blade surface pubescent; anthers 3, ovary with a fleshy appendage below style insertion; inflorescence composed of racemes, fruit pubescent on apex, caryopsis with adherent pericarp.

**Sacred Aspect**

I. In the month Shravana (according to Hindu calendar), the unmarried girls of Vanjara (Banjara) tribe celebrate Tij festival. They raise seedlings of wheat in a pot and donate to the river.

**C.N.:** It is exotic native of Mesopotamia. It is cultivated for food grains.

## 251. *Tridax procumbens* L.

**Family:** Asteraceae

**Local Name:** Ekdandi, Dagdi pala

Decumbent, hispid herbs; leaves opposite, cuneate, ovate-elliptic, inciso-dentate to pinnatifid; flowering head solitary, long peduncled, rays white, disc yellow; achenes obovate or oblong, slightly 3-angled, pappus numerous.

**Sacred Aspect**

I. Herb is found at many sacred places.

**Medicinal Uses**

I. Leaf paste recommended as tooth paste in the treatment on toothache, until cure.

II. Leaf paste is topically applied on cuts and wounds, to avoid infection.

III. Leaf paste is applied on boils and blisters as a coolant.

IV. Root extract is diluted with a cup of warm water. It is prescribed in the treatment on fever. Half cup, thrice a day, recommended for children's for 4-5 days.

**C.N. :** It is exotic, native of Tropical Central America.

## 252. *Vigna mungo* (L.) Hepper

**Family:** Fabaceae

**Local Name:** Udid

Trailing herbs, stem and branches with browns, spreading hairs; leaves 3 foliate, leaflets ovate or rhomboid; flowers papillionaceous, yellow; pods terete, covered with long hairs, seeds black, aril raised.

**Sacred Aspect**

I. Fresh seeds are used as Ranmeva offered to goddess Mahalaxmi.

## 253. *Vigna radiata* (L.) Wilczek

**Family:** Fabaceae

**Local Name:** Sona mug

Twining or suberect herbs, stem slender, angled; leaves 3 foliate; flowers papillionaceous, yellow; pods cylindrical, densely clothed with hairs, seeds greenish yellow.

**Sacred Aspect**

I. Fresh seeds are used as Ranmeva and offered to goddess Mahalaxmi.

## 254. *Vigna trilobata* (L.) Verdc.

**Family:** Fabaceae

**Local Name:** Mung

Prostrate or twining, glabrous or patently hairy herbs; leaves 3-foliate, leaflets ovate, oblong, rhomboid; flowers yellow; pods cylindric, subcompressed, 2-6 seeded, seeds oblong, green or brown with black spot.

**Sacred Aspect**

I. Fresh seeds are used as Ranmeva and offered to goddess Mahalaxmi.

## 255. *Vitex negundo* L.

**Family:** Verbenaceae

**Local Name:** Nirgundi

Large shrubs or small trees; leaves often 3-5 foliate, leaflets lanceolate, acute; flowers small, pale-bluish-purple, branched in racemes; drupes dark purple or black, 4 seeds, ovate-obovate.

**Sacred Aspect**

I. Plant is found at many sacred places.

II. Inflorescence is used in worship of Lord Mhasoba.

**Medicinal Uses**

I. Decoction of leaves is added bath water for paralytic person and also a women after delivery.

II. Leaves are tightly tied on injured part with the help of cloth overnight to treat rheumatism and swelling.

III. Leaf paste is topically applied on boils.

IV. A cup leaf decoction with two teaspoon honey is recommended thrice a day in the treatment of fever, cough and cold.

**C.N.:** It is exotic, native of North China and Mongolia.

## 256. *Vitis vinifera* L.

**Family:** Vitaceae

**Local Name:** Draksha

Extensive, woody vine with bifid tendrils; leaves large, thin, circular or circular-ovate; flowers small, green-berries oval, oblong or globose.

**Sacred Aspect**

I. Fruits are used in various worships.

II. Fruits are the part of food offering to gods.

**Medicinal Uses**

I. Root paste is recommended as toothpaste in toothache treatment.

**C.N.:** It is exotic, native of South-East Europe to West Indies (Singh *et al.*, 2000).

## 257. ***Withania somnifera*** **(L.) Dunal**

**Family:** Solanaceae

**Local Name:** Ashwagandha

Erect white-tomentose, undershrubs, branches terete; leaves simple, acute, ovate; flowers small, greenish- yellow; berries orange-red, globose, seeds numerous, yellow, flat, oblong or spherical and smooth.

**Sacred Aspect**

I. Herb is found at many sacred places.

**Medicinal Uses**

I. Leaf or root paste is topically applied on boils, cuts and wounds.
II. Root decoction with two teaspoon honey is administered in the treatment of cough and fever.

## 258. ***Woodfordia fruticosa*** **(L.) Kurz.**

**Family:** Lythraceae

**Local Name:** Dhayati

Straggling shrubs, bark brown, peeling off in fibres; leaves opposite or subopposite, sessile, ovate-lanceolate; flowers red; capsules oblong or ellipsoid, shining brown, seeds minute, numerous, brown, smooth.

**Sacred Aspect**

I. Plant is found at several sacred places.

**Medicinal Uses**

I. Paste of stem bark is applied on burns, cuts and wounds.
II. Paste of stem powder is topically applied on cuts and wounds.
III. Floral paste is topically applied on the pimples.

## 259. ***Wrightia tinctoria*** **R.Br.**

**Family:** Apocynaceae

**Local Name:** Pandhra kuda

Medium sized trees with milky latex and light greyish-black or ash coloured, bark rough; leaves elliptic-lanceolate or elliptic-oblong; flowers creamy white, in corymbose racemes, follicles linear cylindrical, seeds brown, glabrous.

**Sacred Aspect**

I. Plant is found at many sacred places.

**Medicinal Uses**

I. Bark and root paste is applied on scorpion and other insect bite.

II. Leaf paste is used for mouth wash in case of toothache and throat infection.

## 260. *Xanthium indicum* Koen.

**Family:** Asteraceae

**Local Name:** Sankeshwar, Kutri

Tall, foetid herbs, stem stout; leaves ovate to suborbicular, alternate, often trilobed, serrate; flowering heads globose, green yellow, clothed with hooked spines, achenes oblong-ovoid, flat, black.

**Sacred Aspect**

I. Herb is found at several sacred places.

**Medicinal Uses**

I. Seed oil is topically applied on rheumatism until cure.

II. Root paste is applied topically on burns and boils.

**C.N.:** It is exotic, native of Tropical South America (Patil, 2003).

## 261. *Zea mays* L.

**Family:** Poaceae

**Local Name:** Maka, Makki, Makai.

Monoecious annuals, culms robuss, nodded; leaves lanceolate, broad, flat, apex acute, sheathed; male panicle terminal, female inflorescence axillary; caryopsis subglobose, compressed, large, yellowish-white.

**Sacred Aspect**

I. Maize cobs are worshipped using kumkum and fragrant sticks (Agarbatti) at the time of Duda makay festival celebrated by Pawara tribe.

II. Tribals worship Waghdev and expect that tigers will not trouble the domestic animals. This celebration is carried in jungles. One of the men is decorated as a tiger. The tiger visits the boundary of a village. He collects cobs of maize and other fruits and then cobs are consumed.

**C.N.:** It is exotic, native of Tropical South America. It is cultivated as grain crop in this region.

## 262. *Zingiber officinale* Rosc.

**Family :** Zingibaraceae

**Local Name :** Adrak

Perennial herb, rhizome fleshy, aromatic; leaves distichous, lanceolat to linear-lanceolate, sessile, 10-25 cm long; flowers white, born in spikes.

**Sacred Aspect**

I. The idol of Hanuman, a Hindu god, is worshipped on the day of Hanuman Jayanti on Chaitra Pournima, before sunrise and 'prasad' prepared from dry ginger (*Zinziber officinale* Rosc) is distributed to the devotees.

II. Birthday of Lord Rama, a Hindu god is celebrated on the day of Ram-Navami. 'Prasad' (sweetmeat) prepared using rhizome of this plant is distributed amongst the devotees. Up to 15 cm long, seeds ovoid-oblong, flat, brown, shining.

**Medicinal Uses**

I. Fresh ginger is added in tea to check cough and cold.

**Miscellaneous Uses**

I. It is used as a spice in rarious culinary preparations.

**C.N. :** It is native of tropical Asia, also cultivated in this region and rhizome are sold in local markets.

## 263. *Zizyphus mauritiana* Lam.

**Family:** Rhamnaceae

**Local Name:** Bor

Branched, spiny, large shrubs or small trees, branches downy when young; leaves ovate-elliptic to suorbicular-ovate, rounded at both ends; flowers small, greenish yellow; drupes globose to ovoid, fleshy, smooth, yellow or orange when ripe, stone 1-2 celled.

**Sacred Aspect**

I. Plant is found at many sacred places.

II. Young fruits are used as Ranmeva offered to goddess Mahalaxmi.

III. Sati or Satwai is local diety, worshipped on the fifth day after childbirth. Offering constitute a twig of *Zizyphus mauritiana* Lam., *Clerodendrum phlomidis* L.*f.* and *Curcuma longa* L.

**Medicinal Uses**

I. A teaspoon bark powder is diluted in a glass of water. A pinch of common salt is added. It is advised thrice a day in the treatment of dysentery.
II. Regular eating of fruit is recommended for good digestion.
III. Leaf paste is topically applied on scorpion and other insect bite.
IV. Root paste is topically applied on cuts and wounds.

**C.N.:** It is exotic, native of Tropics and Warm Subtropics (Martin *et al.*, 1987).

# Chapter 5

# Results, Discussion and Conclusions

## Man and Biodiversity

India is esteemed for its rich biodiversity. There are four principal factors which operated in past for our rich endowment of bioresources *viz.* (i) ecosystem diversity, (ii) culinary diversity, (iii) curative diversity and last (iv) cultural diversity (inclusive of spiritual values). Among these, cultural diversity *vis-à-vis* spiritual values appear to have contributed largely to protect and conserve many plants and even animals. Our rich biological heritage has been preserved through ages on account of conservation-oriented religious or cultural traditions. It is the cultural diversity of indigenous human society that modifies or alters the composition and distributive patterns of bioresources through human practices. These practices of production and belief have bearing on biological diversity in vicinity. The relationship of natural resources and mankind *inter se* can be viewed from cultural, moral, ecological and socio-economic boundaries. Man has generally respected these boundaries. Thus biodiversity of a region is inseparable from indigenous culture. Conservation of both the biodiversity and human societies has a symbiotic relationship with each other. To realize the importance of this symbiotic relationship, investigation of the present kind appear relevant, nay a dire necessity of the present time.

## Man and Culture

Man and nature have an intimate relationship *inter se*. Most civilization originated and even flourished in the ambience of plant world. Plants useful to man earned his awe and reverence and their close relationship have been also reflected in temple and holy courtyards. Cultural and social life of a community comprises a wide spectrum of activities, both physical and mental. This is manifested in terms of rites, rituals, customs, traditions, beliefs, faith, religion and traditional wisdom. All these enforce, directly or indirectly, the human society to conserve his bioresources.

## Man and Worship

Man since his appearance on the sylvan Blue Planet is living in nature and also, at the same time recognized its significance. He, therefore, revered nature (including plants and animals) and different planets. He worshiped these since time immemorial. He never thought himself a separate entity from nature and universe. Instead, he considered himself a part of them. He raised temples to worship them. The temples raised by the primitive human societies were simple *e.g.* wood logs or stones.

## Man and his Faith in Plants

Even today, these are observed in many tribal societies. He worshipped trees (or plants). Probably, the earliest form of worship was the veneration of the tree. In our country, plants have long been hold sacred *e.g.* Tulsi (*Ocimum tenuiflorum* L.), Banyan tree (*Ficus benghalensis* L.), Peepal (*Ficus religiosa* L.), Bael [*Aegle marmelos* (L.) Corr.], Ashoka [*Saraca asoca* (Roxb.) de wilde.], *etc.* Before the modern, well-built temples of the present time, trees were open-air shrines (Patil, 2020). There are several other species still awaiting documentation in various Indian societies. The present investigators have made a headway in this direction in Khandesh region, a northern part of Maharashtra State (India). Many species are associated with some legend, beliefs and myths and hence introduced in holy places (Patil, 2020).

## Man in Present Time

Although the various species support delicate network amongst them, in recent times, man has become the major culprit to break this delicate network and symbioism. Because of greed and need, he is polluting water, air and soil and destroying overall environmental fabric and plants (and animals) in particular. In such circumstances, it is imperative to know what is left today. We must be aware about current status of the vegetable kingdom.

## Indian Plant Explorations

Indian subcontinent is rich in biological and cultural diversity. The importance and utility of plants to the people are mentioned in ancient texts and modern literature. However, these inventories are mostly confined to natural vegetation. Biodiversity studies are confined either to some physiographic area or political boundary. Plant explorations whether excursion, exploration or expedition has largely overlooked biodiversity on sacred places wherein man has introduced many species.

## Studies in the Region

Hooker and Thompson (1883) remarked Khandesh region (northern region of Maharashtra State) a botanically distinct subprovince. It consists presently of three districts *viz*; Jalgaon, Dhule and Nandurbar districts. Jalgaon district is floristically investigated by Kshirsagar and Patil (2008), whereas Dhule and Nandurbar districts by Patil (2003). Jalgaon district is studied from ethnobotanical point of view (Pawar and Patil 2008), however, Dhule and Nandurbar district received a little attention on this line. The literary resume indicated that the study on sacred plants needs a special attention to give full justice to the subject matter. Plant species have an omnipresent role in society w.r.t. medicine, miscellaneous utilities, rituals, rites. Since plants often possesses significant cultural values for society, this kind of information might amplify our understanding of their presence in holy places or temples. In order to understand their social value, the present inventory was intended. It is, therefore, at this backdrop, studied as an exclusive topic of research, the results of which are dilated in the following.

## People in the Region and available Bioresources

This region is basically agrarian. Forested areas extend on the north-western side of the region as Satpura Mountain and few ranges of Western Ghats. The forests, in general, are dry deciduous type. Majority of people belong to Hindu community, apart from fair representation of other religions like Muslims and Buddha. Apart from urban elites, the region is predominated by rurals folks and tribes such as Bhil, Pawara, Kokani, Mavachi, *etc.* The tribals have their own distinctive way of life of their own. Obviously, all these people observe their own religion, rites, rituals and customs in respect of their gods, goddesses, local deities (gram-devatas), festivals, marriages, births and such other social activities. They offer or use one or other plant species or their parts/plant products. These can be analyzed as the following.

The region under study, being mainly agrarian, cultivates some crops belonging to pulses (cowpea, mungbean, blackgram, pigeon pea, chicken pea, horse gram *etc.*), cereals (maize, wheat, rice), millets (sorghum, pearl, millet, kodra, banti, foxtail millet, *etc.*), pseudo-cereals (*Amaranthus* species.), vegetables, fruit yielding and economically important crops like sugarcane and cotton. The wild plants are also fairly represented in this part of Maharashtra *e.g.* 1333 species in Dhule and Nandurbar district(Patil, 2003) and 813 species in Jalgaon district (Kshirsagar and Patil, 2008). Thus area under study has rich plant diversity.

## Floristic Analysis

The present investigators revealed fair diversity of plants used for sacred or religious purposes. As many as 263 species belonging to 221 genera and 83 families are noted useful, one being gymnospermic (Table 1). However, no indigenous species of gymnosperms inhabit in this region. Species of *Thuja* is usually found under plantation. The lower group of plants *e.g.* pteridophytes are also poorly represented in this region. Of these, major segment of plant diversity useful for the various sacred or religious purposes in the region belong to dicotyledons. They are represented by total of 232 species belonging to 194 genera and 73 families. Monocotyledons are represented by 31 species belonging to 27 genera and 10 families.

## Habital Groups

The total 263 species belong to different habital groups of plants. These are trees (89 species), shrubs (49 species), climbers (30 species) and herbs (96 species). Out of 89 species of trees, only 04 species are monocotyledons and the rest others (84 species) are dicotyledons. Thus the arborescent dicotyledons share a major segment of useful species. Overall, herbaceous floral elements, weather monocotyledons or dicotyledons, are employed on a large scale (Table 1).

**Table 1: Habitat Groups of Species Recorded**

| Sl. No. | Botanical Name | Family | Herb | Shrub | Climber/ Lianas | Trees | Geo-phytes |
|---|---|---|---|---|---|---|---|
| 1. | *Abelmoschus esculentus* (L.) Moench. | Malvaceae | √ | | | | |
| 2. | *Abelmoschus manihot* (L.) Medik. | Malvaceae | √ | | | | |
| 3. | *Abitulon indicum* (L.) | Malvaceae | √ | | | | |
| 4. | *Abrus precatorius* L. | Fabaceae | | | √ | | |
| 5. | *Acacia chundra* (Roxb. ex Rottle.) Willd. | Mimosaceae | | √ | | | |
| 6. | *Acacia ferruginea* DC. | Mimosaceae | | | | √ | |
| 7. | *Acacia nilotica* (L.) Del. | Mimosaceae | | | | √ | |
| 8. | *Acalypha indica* L. | Euphorbiaceae | √ | | | | |
| 9. | *Achyranthes aspera* L. | Amarantaceae | √ | | | | |
| 10. | *Acorus calamus* L. | Araceae | √ | | | | |
| 11. | *Adansonia digitata* L. | Bombacaceae | | | | √ | |
| 12. | *Aegle marmelos* (L.) Corr. | Rutaceae | | | | √ | |
| 13. | *Ailanthus excelsa* Roxb. | Simarubaceae | | | | √ | |
| 14. | *Albizia lebbeck* (L.) Willd. | Mimosaceae | | | | √ | |
| 15. | *Allium cepa* L. | Liliaceae | √ | | | | |
| 16. | *Allium sativum* L. | Liliaceae | √ | | | | |

| Sl. No. | Botanical Name | Family | Herb | Shrub | Climber/ Lianas | Trees | Geo-phytes |
|---|---|---|---|---|---|---|---|
| 17. | *Aloe vera* L. | Liliaceae | √ | | | | |
| 18. | *Alstonia scholaris* R.Br. | Apocynaceae | | | | √ | |
| 19. | *Amaranthus hybridus* L | Amaranthaceae | √ | | | | |
| 20. | *Amaranthus spinosus* L. | Amaranthaceae | √ | | | | |
| 21. | *Ammannia baccifera* L. | Lythraceae | √ | | | | |
| 22. | *Andrographis echioides* (L.) Nees. | Acanthaceae | √ | | | | |
| 23. | *Anethum graveolens* L. | Labiatae | √ | | | | |
| 24. | *Anisomeles malabarica* R. Br. | Annonaceae | | √ | | | |
| 25. | *Annona reticulata* L. | Umbellifereae | | | | √ | |
| 26. | *Annona squamosa* L. | Annonaceae | | √ | | | |
| 27. | *Anogeissus latifolia* Wall. | Combretaceae | | | | √ | |
| 28. | *Apium graveolens* L. | Apiaceae | √ | | | | |
| 29. | *Arachis hypogaea* L. | Fabaceae | √ | | | | |
| 30. | *Argemone mexicana* L. | Papaveranceae | √ | | | | |
| 31. | *Argyreia nervosa* (Burm. *f.*) Bojer | Convolvulaceae | | | √ | | |
| 32. | *Aristolochia bracteolata* Lam. | Aristolochiaceae | √ | | | | |
| 33. | *Asparagus racemosus* Willd | Liliaceae | √ | | | | |
| 34. | *Azadirachta indica* A. Juss. | Meliaceae | | | | √ | |
| 35. | *Bacopa monnieri* Penn. | Scrophularia-ceae | √ | | | | |
| 36. | *Balanitis aegyptiaca* (L.) Delile | Balanitaceae | | √ | | | |
| 37. | *Baliospermum raziana* Keshav *et.* Yog | Euphorbiaceae | √ | | | | |
| 38. | *Bambusa arundinacea* (Retz.)Willd. | Poaceae | | | | √ | |
| 39. | *Barleria lawii* L. | Acanthaceae | | √ | | | |
| 40. | *Barleria prionitis* L. | Acanthaceae | | √ | | | |
| 41. | *Bauhinia racemosa* Lamk. | Caesalpiniaceae | | | | √ | |
| 42. | *Bauhinia tomentosa* L. | Caesalpiniaceae | | | | √ | |
| 43. | *Bauhinia variegata* L. | Caesalpiniaceae | | | | √ | |
| 44. | *Benincasa hispida* (Thunb.) Cogn | Cucurbitaceae | | | √ | | |
| 45. | *Boerhavia repens* L. var. *diffusa* (L.) Hook *f.* | Nyctaginaceae | √ | | | | |

| *Sl. No.* | *Botanical Name* | *Family* | *Herb* | *Shrub* | *Climber/ Lianas* | *Trees* | *Geo-phytes* |
|---|---|---|---|---|---|---|---|
| 46. | *Bombax ceiba* L. | Bombacaceae | | | | √ | |
| 47. | *Borassus flabellifer* L. | Aracaceae | | | | √ | |
| 48. | *Boswellia serrata* Roxb. | Burseraceae | | | | √ | |
| 49. | *Bougainvillea spectabilis* Willd. | Nyctaginaceae | | √ | | | |
| 50. | *Bridelia retusa* (L.) Spreng. | Euphorbiaceae | | | | √ | |
| 51. | *Buchanania lanzan* Spreng. | Anacardiaceae | | | | √ | |
| 52. | *Butea monosperma* (Lam.) Taub. | Fabaceae | | | | √ | |
| 53. | *Caesalpinia bonduc* (L.) Roxb. | Caesalpiniaceae | | | | √ | |
| 54. | *Caesalpinia pulcherrima* (L.) Sw | Caesalpiniaceae | | √ | | | |
| 55. | *Cajanus cajan* (L.) Millsp. | Papilionaceae | | √ | | | |
| 56. | *Calotropis gigantea* (L.) R. Br. | Asclepiadaceae | | √ | | | |
| 57. | *Calotropis procera* (Ait.) R. Br. | Asclepiadaceae | | √ | | | |
| 58. | *Canna indica* L. | Cannaceae | √ | | | | |
| 59. | *Cardiospermum halicacabum* L. | Sapindaceae | | | √ | | |
| 60. | *Carica papaya* L. | Caricaceae | | | | √ | |
| 61. | *Cascabella thevetia* (L.) Lippold | Apocynaceae | | √ | | | |
| 62. | *Cassia absus* L. | Caesalpiniaceae | √ | | | | |
| 63. | *Cassia auriculata* L | Caesalpiniaceae | | √ | | | |
| 64. | *Cassia fistula* L. | Caesalpiniaceae | | | | √ | |
| 65. | *Cassia occidentalis* L. | Caesalpiniaceae | | √ | | | |
| 66. | *Cassia sophera* L. | Caesalpiniaceae | | √ | | | |
| 67. | *Cassia tora* L. | Caesalpiniaceae | √ | | | | |
| 68. | *Catharanthus roseus* (L.) G.Don | Apocynaceae | √ | | | | |
| 69. | *Celastrus paniculata* Willd. | Celatraceae | | √ | | | |
| 70. | *Celosia argentea* L. | Amranthaceaae | √ | | | | |
| 71. | *Centella asiatica* L. | Apiaceae | √ | | | | |
| 72. | *Chlorophytum borivilianum* Sant. and Fernand. | Liliaceae | √ | | | | |

| Sl. No. | Botanical Name | Family | Herb | Shrub | Climber/ Lianas | Trees | Geo-phytes |
|---|---|---|---|---|---|---|---|
| 73. | *Chrysanthemum indicum* L. | Asteraceae | √ | | | | |
| 74. | *Cicer arietinum* L. | Fabaceae | √ | | | | |
| 75. | *Cissampelos pareira* L. | Menispermaceae | | | √ | | |
| 76. | *Cissus quadrangularis* L | Vitaceae | | | √ | | |
| 77. | *Citrullus colocynthis* Sch. | Cucurbitaceae | | | √ | | |
| 78. | *Citrullus lanatus* (Thunb.) Matsura and Nakai | Cucurbitaceae | | | √ | | |
| 79. | *Citrus aurantifolia* (Christm. and Panz.) Sw. | Rutaceae | | | | √ | |
| 80. | *Citrus aurantium* L. | Rutaceae | | | | √ | |
| 81. | *Cleome gynandra* L | Capparidaceae | √ | | | | |
| 82. | *Cleome viscosa* L | Capparidaceae | √ | | | | |
| 83. | *Clitoria biflora* Dalz. | Fabaceae | √ | | | | |
| 84. | *Clitoria ternatea* L. | Fabaceae | | | √ | | |
| 85. | *Coccinia grandis* (L.) Voigt | Cucurbitaceae | | | √ | | |
| 86. | *Cocculus hirsutus* (L.) Theob. | Menispermaceae | | | √ | | |
| 87. | *Cocos nucifera* L. | Arecaceae | | | | √ | |
| 88. | *Corchorus capsularis* L. | Tiliaceae | √ | | | | |
| 89. | *Cordia dichotoma* Forst. *f.* | Boraginaceae | | | | √ | |
| 90. | *Coriandrum sativum* L. | Umbellifereae | √ | | | | |
| 91. | *Costus specious* (Koen.) J. E. Sm. | Zingiberaceae | | √ | | | |
| 92. | *Crataeva adansonii* DC. | Capparidaceae | | | | √ | |
| 93. | *Crinum asiaticum* L. | Amaryllidaceae | √ | | | | |
| 94. | *Cucumis melo* L. | Cucurbitaceae | | | √ | | |
| 95. | *Cucumis sativus* L. | Cucurbitaceae | | | √ | | |
| 96. | *Cucurbita maxima* Duch. ex Lam. | Cucurbitaceae | | | √ | | |
| 97. | *Curculigo orchioides* Gaertn. | Hypoxidaceae | √ | | | | |
| 98. | *Curcuma amada* Roxb. | Zingiberaceae | √ | | | | |
| 99. | *Curcuma inodora* Blatt. | Zingiberaceae | √ | | | | |
| 100. | *Curcuma longa* L. | Zingiberaceae | √ | | | | |
| 101. | *Cymbopogon citratus* (DC.) Stapf | Poaceae | √ | | | | |
| 102. | *Cymbopogon martinii* (Roxb.) Wats. | Poaceae | √ | | | | |

| Sl. No. | Botanical Name | Family | Herb | Shrub | Climber/ Lianas | Trees | Geo-phytes |
|---|---|---|---|---|---|---|---|
| 103. | *Cynodon dactylon* (L.) Pers. | Poaceae | √ | | | | |
| 104. | *Cyperus rotundus* L | Cyperaceae | √ | | | | |
| 105. | *Dalbergia sissoo* Roxb. ex DC. | Fabaceae | | | | √ | |
| 106. | *Datura innoxia* Mill. | Solanaceae | √ | | | | |
| 107. | *Datura metel* L. | Solanaceae | √ | | | | |
| 108. | *Daucus carota* L. | Apiaceae | √ | | | | |
| 109. | *Delonix regia* (Boj. ex Hook.) Raf. | Caesalpiniaceae | | | | √ | |
| 110. | *Desmodium gangeticum* (L.) DC. | Fabaceae | | √ | | | |
| 111. | *Diospyros melanoxylon* Roxb | Ebenaceae | | | | √ | |
| 112. | *Diospyros montana* Roxb | Ebenaceae | | | | √ | |
| 113. | *Diplocyclos palmatus* (L.) Jeffrey | Cucurbitaceae | | | √ | | |
| 114. | *Dolichandrone falcata* (Wall.ex DC) Seem. | Bignoniaceae | | | | √ | |
| 115. | *Drimia indica* (Roxb.) Jessop | Liliaceae | √ | | | | |
| 116. | *Echinops echinatus* Roxb. | Asteraceae | √ | | | | |
| 117. | *Eclipta prostrata* (L.) L. | Asteraceae | √ | | | | |
| 118. | *Elaeocarpus sphaericus* Gaertn. | Elaeaocarpa-ceae | | | | √ | |
| 119. | *Eleusine coracana* (L.) Gaertn. | Poaceae | √ | | | | |
| 120. | *Emblica officinalis* Gaertn. | Euphorbiaceae | | | | √ | |
| 121. | *Enicostema axillare* (Lam.) Raynal | Gentianaceae | √ | | | | |
| 122. | *Euphorbia hirta* L. | Euphorbiaceae | √ | | | | |
| 123. | *Evolvulus alsinoides* (L.) L. | Convolvulaceae | √ | | | | |
| 124. | *Ficus benghalensis* L. | Moraceae | | | | √ | |
| 125. | *Ficus racemosa* L. | Moraceae | | | | √ | |
| 126. | *Ficus religiosa* L. | Moraceae | | | | √ | |
| 127. | *Ficus virens* Ait. | Moraceae | | | | √ | |
| 128. | *Flacourtia indica* (Burm. f.) Merr. | Flacourtiaceae | | | | √ | |
| 129. | *Garnga pinnata* Roxb. | Burseraceae | | | | √ | |
| 130. | *Gloriosa superba* L. | Liliaceae | | | √ | | |

| Sl. No. | Botanical Name | Family | Herb | Shrub | Climber/ Lianas | Trees | Geo-phytes |
|---|---|---|---|---|---|---|---|
| 131. | *Gmelina arborea* Roxb. | Verbenaceae | | | | √ | |
| 132. | *Gossypium arboreum* L. | Malvaceae | | √ | | | |
| 133. | *Gossypium herbaceum* L. | Malvaceae | | √ | | | |
| 134. | *Haldina cordifolia* (Roxb.) Ridsd. | Rubiaceae | | | | √ | |
| 135. | *Hardwickia binata* Roxb. | Caesalpiniaceae | | | | √ | |
| 136. | *Helicteres isora* L. | Sterculiaceae | | √ | | | |
| 137. | *Hemidesmus indicus* (L.) Schult. | Asclepiadaceae | | | √ | | |
| 138. | *Hibiscus rosa-sinensis* L. | Malvaceae | | √ | | | |
| 139. | *Hibiscus sabdariffa* L. | Malvaceae | √ | | | | |
| 140. | *Holarrhena pubescens* (Buch.-Ham.) Wall. ex G. Don | Apocynaceae | | | | √ | |
| 141. | *Holoptelea integrifolia* Planch. | Ulmaceae | | | | √ | |
| 142. | *Hymenodictyon ovixense* (Roxb.) Mabberly | Rubiaceae | | | | √ | |
| 143. | *Ipomoea illustris* (Clarke) Prain. | Convolvulaceae | | | √ | | |
| 144. | *Ipomoea nil* (L.) Roth | Convolvulaceae | √ | | | | |
| 145. | *Ixora coccinea* L. | Rubiaceae | | √ | | | |
| 146. | *Jasminum officinale* L. | Oleaceae | | √ | | | |
| 147. | *Jasminum sambac* (L.) Ait. | Oleaceae | | √ | | | |
| 148. | *Jatropha curcas* L. | Euphorbiaceae | | √ | | | |
| 149. | *Justicia adhatoda* L. | Acanthaceae | | √ | | | |
| 150. | *Kalanchoe pinnata* (Lam.) Pers. | Crassulaceae | √ | | | | |
| 151. | *Kohautia aspera* (Heyne ex Roth) Brem. | Rubiaceae | √ | | | | |
| 152. | *Lagenaria siceraria* (Mol.) Standl. | Cucurbitaceae | | | √ | | |
| 153. | *Lagerstroemia microcarpa* Wight | Lythraceae | | | | √ | |
| 154. | *Lantana camara* L. | Verbenaceae | | √ | | | |
| 155. | *Lawsonia inermis* L. | Lythraceae | | √ | | | |
| 156. | *Leucas cephalotes* (Roth) Spr. | Lamiaceae | √ | | | | |
| 157. | *Limonia acidissima* L. | Rutaceae | | | | √ | |
| 158. | *Luffa acutangula* (L.) Roxb. | Cucurbitaceae | | | √ | | |

| Sl. No. | Botanical Name | Family | Herb | Shrub | Climber/ Lianas | Trees | Geo-phytes |
|---|---|---|---|---|---|---|---|
| 159. | *Madhuca longifolia* (Koen.) Mac Bride | Sapotaceae | | | | √ | |
| 160. | *Mallotus philippensis* (Lam.) Muell-Arg. | Euphorbiaceae | | | | √ | |
| 161. | *Mangifera indica* L. | Anacardiaceae | | | | √ | |
| 162. | *Martynia annua* L. | Martyniaceae | √ | | | | |
| 163. | *Melia azedarach* L. | Meliaceae | | | | √ | |
| 164. | *Melilotus alba* Medik. ex Desr. | Fabaceae | √ | | | | |
| 165. | *Merremia dissecta* (Jacq.) Hallier *f.* | Convolvulaceae | √ | | | | |
| 166. | *Michelia champaca* L. | Magnoliaceae | | | | √ | |
| 167. | *Mimosa hamata* Willed. | Mimosaceae | | √ | | | |
| 168. | *Mimusops elengi* L. | Sapotaceae | | | | √ | |
| 169. | *Mirabilis jalapa* L. | Nyctaginaceae | √ | | | | |
| 170. | *Mitragyna parvifolia* (Roxb.) Korth. | Rubiaceae | | | | √ | |
| 171. | *Momordica dioica* Roxb. ex Willd. | Cucurbitaceae | | | √ | | |
| 172. | *Moringa oleifera* Lam. | Moringaceae | | | | √ | |
| 173. | *Mucuna pruriens* (L.) DC. | Fabaceae | | √ | | | |
| 174. | *Mukia maderaspatana* L. | Cucurbitaceae | | | √ | | |
| 175. | *Musa paradisiaca* L. | Musaceae | | | | √ | |
| 176. | *Nelumbo nucifera* Gaertn. | Nelumbonaceae | √ | | | | |
| 177. | *Nerium indicum* Mill. | Apocynaceae | | √ | | | |
| 178. | *Nyctanthes arbortristis* L. | Oleaceae | | | | √ | |
| 179. | *Nymphaea nouchali* Burm.*f.* | Nymphaeaceae | √ | | | | |
| 180. | *Nymphaea pubescens* Willd | Nymphaeaceae | √ | | | | |
| 181. | *Ocimum basilicum* L. | Lamiaceae | √ | | | | |
| 182. | *Ocimum gratissimum* L. | Lamiaceae | | √ | | | |
| 183. | *Ocimum tenuiflorum L.* | Lamiaceae | | √ | | | |
| 184. | *Opuntia elatior* Mill. | Cactaceae | | √ | | | |
| 185. | *Oroxylum indicum* (L.) Vent. | Bignoniaceae | | | | √ | |
| 186. | *Oryza sativa* L. | Poaceae | √ | | | | |
| 187. | *Oxalis corniculata* L. | Oxalidaceae | √ | | | | |
| 188. | *Pandanus odoratissimus* L. *f.* | Pandanaceae | | | | √ | |
| 189. | *Passiflora edulis* Sims | Passifloraceae | | | √ | | |

| Sl. No. | Botanical Name | Family | Herb | Shrub | Climber/ Lianas | Trees | Geo-phytes |
|---|---|---|---|---|---|---|---|
| 190. | *Pedalium murex* L. | Pedaliaceae | √ | | | | |
| 191. | *Pennisetum americanum* L. | Poaceae | √ | | | | |
| 192. | *Pergularia daemia* (Forssk.) Chiov. | Asclepiadoideae | | | √ | | |
| 193. | *Phoenix dactylifera* (L) Roxb. | Arecaceae | | | | √ | |
| 194. | *Phyla nodiflora* (L.) Greene | Verbenaceae | √ | | | | |
| 195. | *Phyllanthus acidus* (L.) K. Skeels | Euphorbiaceae | | | | √ | |
| 196. | *Phyllanthus amarus* L. | Euphorbiaceae | √ | | | | |
| 197. | *Physalis micrantha* L. | Solanaceae | √ | | | | |
| 198. | *Piper betle* L. | Piperaceae | | √ | | | |
| 199. | *Plumbago zeylanica* L. | Plumbaginaceae | | √ | | | |
| 200. | *Plumeria rubra* L. forma *acuminata* (Ait.) Santapaul and Irani ex Shah | Apocynaceae | | | | √ | |
| 201. | *Polyalthia longifolia* Sonn. | Annonaceae | | | | √ | |
| 202. | *Pongamia pinnata* (L.) Pierre | Lamiaceae | | | | √ | |
| 203. | *Prosopis cineraria* (L.) Druce | Mimosaceae | | | | √ | |
| 204. | *Psidium guajava* L. | Myrtaceae | | | | √ | |
| 205. | *Pterocarpus marsupium* Roxb. | Fabaceae | | | | √ | |
| 206. | *Pueraria tuberosa* (Roxb. ex Willd.) DC. | Fabaceae | | | √ | | |
| 207. | *Punica granatum* L. | Punicaceae | | √ | | | |
| 208. | *Quisqualis indica* L. | Combretaceae | | | √ | | |
| 209. | *Raphanus sativus* L. | Brassicaceae | √ | | | | |
| 210. | *Ricinus communis* L. | Euphorbiaceae | | √ | | | |
| 211. | *Rosa damascena* Mill. | Rosaceae | | √ | | | |
| 212. | *Saccharum officinarum* L. | Poaceae | | √ | | | |
| 213. | *Salvadora persica* L. | Salvadoraceae | | | | √ | |
| 214. | *Santalum album* L. | Santalaceae | | | | √ | |
| 215. | *Sapindus emarginatus* Vahl. | Sapindaceae | | | | √ | |
| 216. | *Saraca asoca* (Roxb.) de Wilde | Caesalpiniaceae | | | | √ | |

| Sl. No. | Botanical Name | Family | Herb | Shrub | Climber/ Lianas | Trees | Geo- phytes |
|---|---|---|---|---|---|---|---|
| 217. | *Schleichera oleosa* (Lour.) Oken | Sapindaceae | | | | √ | |
| 218. | *Securinega virosa* (Roxb. ex Willd.) Baill. | Euphorbiaceae | | √ | | | |
| 219. | *Semecarpus anacardium* L. | Anacardiaceae | | | | √ | |
| 220. | *Sesamum orientale* L. | Pedaliaceae | √ | | | | |
| 221. | *Sesbania grandiflora* (L.) Poir. | Fabaceae | | | | √ | |
| 222. | *Sesbania sesban* (L.) Merr. | Fabaceae | | | | √ | |
| 223. | *Setaria italica* (L.) P. Beauv. | Poaceae | √ | | | | |
| 224. | *Sida acuta* Burm. *f.* | Malvaceae | | √ | | | |
| 225. | *Sida cordata* (Burm.*f.*) Borss. | Malvaceae | √ | | | | |
| 226. | *Sida spinosa* L. | Malvaceae | √ | | | | |
| 227. | *Solanum anguivi* Lam. | Solanaceae | √ | | | | |
| 228. | *Solanum melongena* L. | Solanaceae | √ | | | | |
| 229. | *Solanum nigrum* L. | Solanaceae | | √ | | | |
| 230. | *Solanum virginianum* L. | Solanaceae | √ | | | | |
| 231. | *Sorghum vulgare* Pers. | Poaceae | √ | | | | |
| 232. | *Spilanthus calva* DC. | Asteraceae | √ | | | | |
| 233. | *Spondias pinnata* (L. *f.*) Kurz. | Anacardiaceae | | | | √ | |
| 234. | *Sterculia urens* Roxb. | Sterculiaceae | | | | √ | |
| 235. | *Syzygium cumini* (L.) Skeels | Myrtaceae | | | | √ | |
| 236. | *Tabernaemontana divaricata* (L.) R. Br. | Apocynaceae | | √ | | | |
| 237. | *Tagetes erecta* L. | Asteraceae | √ | | | | |
| 238. | *Tamarindus indica* L. | Caesalpiniaceae | | | | √ | |
| 239. | *Tectona grandis* L. | Verbenaceae | | | | √ | |
| 240. | *Tephrosia purpurea* Pers. | Fabaceae | | √ | | | |
| 241. | *Terminalia arjuna* (Roxb. ex DC.) Wight and Arn. | Combretaceae | | | | √ | |
| 242. | *Terminalia bellirica* (Gaertn.) Roxb. | Combretaceae | | | | √ | |
| 243. | *Thespesia populnea* (L.) Soland. ex Corr. | Malvaceae | | | | √ | |
| 244. | *Thuja plicata* ex D. Don | Cupressaceae | | | | √ | |

| Sl. No. | Botanical Name | Family | Herb | Shrub | Climber/ Lianas | Trees | Geo-phytes |
|---|---|---|---|---|---|---|---|
| 245. | *Tinospora cordifolia* (Willd.) Miers ex Hook. *f.* and Thoms | Menisperma-ceae | | | √ | | |
| 246. | *Tribulus terrestris* L. | Zygophyllaceae | √ | | | | |
| 247. | *Tricholepis indica* (Willd.) Almeida | Asteraceae | √ | | | | |
| 248. | *Trichosanthes cucumerina* L. | Cucurbitaceae | | | √ | | |
| 249. | *Trichosanthes tricuspidata* Lour. | Cucurbitaceae | | | √ | | |
| 250. | *Tricticum aestivum* L. | Poaceae | √ | | | | |
| 251. | *Tridax procumbens* L. | Asteraceae | √ | | | | |
| 252. | *Vigna mungo* (L.) Hepper | Fabaceae | √ | | | | |
| 253. | *Vigna radiata* (L.) R. Wilczek | Fabaceae | √ | | | | |
| 254. | *Vigna trilobata* (L.) Verdc. | Fabaceae | √ | | | | |
| 255. | *Vitex negundo* L. | Verbenaceae | | | | √ | |
| 256. | *Vitis vinifera* L. | Vitaceae | | | √ | | |
| 257. | *Withania somnifera* (L.) Dunal | Solanaceae | | √ | | | |
| 258. | *Woodfordia fruticosa* (L.) Kurz. | Lythraceae | | √ | | | |
| 259. | *Wrightia tinctoria* R.Br. | Apocynaceae | | | | √ | |
| 260. | *Xanthium indicum* Koen. | Asteraceae | √ | | | | |
| 261. | *Zea mays* L. | Poaceae | √ | | | | |
| 262. | *Zingiber officinale* Rosc. | Zingibaraceae | √ | | | | |
| 263. | *Zizyphus mauritiana* Lam. | Rhamnaceae | | √ | | | |

## Status Regarding Cultivated Species

Out of the total 263 species, 103 species are already found under cultivation (Table 2). These are locally and traditionally used for various miscellaneous purposes such as food crops (pulses, cereals, millets, *etc.*), edible fruits, vegetables, spices and condiments, ornamentals or garden plants, timber, oil-yielding, fiber-yielders, *etc.* The indigenous communities appear to have appropriated them for religious and sacred purposes. These are being used by the members of all indigenous human societies in this region. A fair number of cultigens thus found place in worships, rites, rituals and various social ceremonies. Such species need no special thinking as they are part of life-supporting system locally and thus sustain local populace on a large scale. Only 18 species of cultivated ones are being exhausted for these purposes. Again, a major share of utility is hailed from the dicotyledonous taxa. It seems that use

**Table 2: Status of Species Recorded**

| *Sl. No.* | *Botanical Name* | *Local Name* | *Family* | *Exotic* | *Indi-genous* | *Wild* | *Culti-vated* | *Cultivated and Wild Both* |
|---|---|---|---|---|---|---|---|---|
| 1. | *Abelmoschus esculentus* (L.) Moench. | Bhendi | Malvaceae | √ | | | √ | |
| 2. | *Abelmoschus manihot* (L.) Medik. | Kasturi bhendi, Ran bhendi | Malvaceae | | √ | √ | | |
| 3. | *Abitulon indicum* (L.) | Sikka, Atti. | Malvaceae | | √ | √ | | |
| 4. | *Abrus precatorius* L. | Gunj, Ratti | Fabaceae | | √ | √ | | |
| 5. | *Acacia chundra* (Roxb. ex Rottle.) Willd. | Khair | Mimosaceae | | √ | √ | | |
| 6. | *Acacia ferruginea* DC. | Devbhabul | Mimosaceae | | √ | √ | | |
| 7. | *Acacia nilotica* (L.) Del. | Ramkathi. | Mimosaceae | | √ | √ | | |
| 8. | *Acalypha indica* L. | Kokali | Euphorbiaceae | | √ | √ | | |
| 9. | *Achyranthes aspera* L. | Aghada, Chirchir | Amarantaceae | | √ | √ | | |
| 10. | *Acorus calamus* L. | Vekhand; Yekhand | Araceae | | √ | √ | | |
| 11. | *Adansonia digitata* L. | Gorakh chinch | Bombacaceae | √ | | √ | | |
| 12. | *Aegle marmelos* (L.) Corr. | Bel phal, Bel mool, Bel. | Rutaceae | | √ | | | √ |
| 13. | *Ailanthus excelsa* Roxb. | Maharuk | Simarubaceae | | √ | √ | | |
| 14. | *Albizia lebbeck* (L.) Willd. | Shirish | Mimosaceae | √ | | √ | | |
| 15. | *Allium cepa* L. | Kanda | Liliaceae | √ | | | √ | |
| 16. | *Allium sativum* L. | Lasun | Liliaceae | √ | | | √ | |
| 17. | *Aloe vera* L. | Korphad | Liliaceae | √ | | | | √ |
| 18. | *Alstonia scholaris* R.Br. | Saptaparn | Apocynaceae | | √ | | √ | |
| 19. | *Amaranthus hybridus* L | Rajgira | Amaranthaceae | | √ | √ | | |
| 20. | *Amaranthus spinosus* L. | Kate – math | Amaranthaceae | √ | | √ | | |

| Sl. No. | Botanical Name | Local Name | Family | Exotic | Indi-genous | Wild | Culti-vated | Cultivated and Wild Both |
|---|---|---|---|---|---|---|---|---|
| 21. | *Ammannia baccifera* L. | Aagia | Lythraceae | | √ | √ | | |
| 22. | *Andrographis echioides* (L.) Nees. | Kalmegh | Acanthaceae | | √ | √ | | |
| 23. | *Anethum graveolens* L. | Kalatumba | Labiatae | | √ | √ | | |
| 24. | *Anisomeles malabarica* R. Br. | Ramphal. | Annonaceae | √ | | | √ | |
| 25. | *Annona reticulata* L. | Balant shepu | Umbellifereae | √ | | | √ | |
| 26. | *Annona squamosa* L. | Sitaphal | Annonaceae | √ | | | √ | |
| 27. | *Anogeissus latifolia* Wall. | Dhavda | Combretaceae | | √ | √ | | |
| 28. | *Apium graveolens* L. | Shepu | Apiaceae | | √ | | √ | |
| 29. | *Arachis hypogaea* L. | Bhuimug | Fabaceae | | √ | | √ | |
| 30. | *Argemone mexicana* L. | Pivla Dhotra | Papaveranceae | √ | | √ | | |
| 31. | *Argyreia nervosa* (Burm. *f.*) Bojer | Samudrashok | Convolvulaceae | | √ | | √ | |
| 32. | *Aristolochia bracteolata* Lam. | Kidamar | Aristolochiaceae | | √ | √ | | |
| 33. | *Asparagus racemosus* Willd | Shatavari, Chinyar | Liliaceae | | √ | | | √ |
| 34. | *Azadirachta indica* A. Juss. | Kadu neem | Meliaceae | | √ | | | √ |
| 35. | *Bacopa monnieri* Penn. | Brahmi | Scrophulariaceae | | √ | | | √ |
| 36. | *Balanitis aegyptiaca* (L.) Delile | Hinganbet | Balanitaceae | | √ | √ | | |
| 37. | *Baliospermum raziana* Keshav *et.* Yog | Danti | Euphorbiaceae | | √ | √ | | |
| 38. | *Bambusa arundinacea* (Retz.)Willd. | Kalak; Bambu. | Poaceae | | √ | √ | | |
| 39. | *Barleria lawii* L. | Akhar | Acanthaceae | | √ | √ | | |
| 40. | *Barleria prionitis* L. | Kate Koranti. | Acanthaceae | | √ | | | √ |

| Sl. No. | Botanical Name | Local Name | Family | Exotic | Indi-genous | Wild | Culti-vated | Cultivated and Wild Both |
|---|---|---|---|---|---|---|---|---|
| 41. | *Bauhinia racemosa* Lamk. | Apta | Caesalpiniaceae | | √ | √ | | |
| 42. | *Bauhinia tomentosa* L. | Ashundar | Caesalpiniaceae | | √ | √ | | |
| 43. | *Bauhinia variegata* L. | Kanchan. | Caesalpiniaceae | | √ | | √ | |
| 44. | *Benincasa hispida* (Thunb.) Cogn | Kohla. | Cucurbitaceae | √ | | | √ | |
| 45. | *Boerhavia repens* L. var. *diffusa* (L.) Hook *f.* | Punarnawa | Nyctaginaceae | | √ | √ | | |
| 46. | *Bombax ceiba* L. | Kate Sawar, Hawar | Bombacaceae | | √ | √ | | |
| 47. | *Borassus flabellifer* L. | Tad | Aracaceae | √ | | | | √ |
| 48. | *Boswellia serrata* Roxb. | Salayi. | Burseraceae | | √ | √ | | |
| 49. | *Bougainvillea spectabilis* Willd. | Bogan-wel | Nyctaginaceae | √ | | | √ | |
| 50. | *Bridelia retusa* (L.) Spreng. | Asana | Euphorbiaceae | | √ | √ | | |
| 51. | *Buchanania lanzan* Spreng. | Charoli. | Anacardiaceae | | √ | | | √ |
| 52. | *Butea monosperma* (Lam.) Taub. | Palas, Dhak | Fabaceae | | √ | √ | | |
| 53. | *Caesalpinia bonduc* (L.) Roxb. | Sagar-gota | Caesalpiniaceae | | √ | √ | | |
| 54. | *Caesalpinia pulcherrima* (L.) Sw. | Puraiful | Caesalpiniaceae | √ | | √ | | |
| 55. | *Cajanus cajan* (L.) Millsp. | Tur; Tor. | Papilionaceae | | √ | | √ | |
| 56. | *Calotropis gigantea* (L.) R. Br | Rui; Pandhari mandar; Arka mool | Asclepiadaceae | | √ | √ | | |
| 57. | *Calotropis procera* (Ait.) R. Br. | Rui; Arka mool | Asclepiadaceae | | √ | √ | | |
| 58. | *Canna indica* L. | Kardali. | Cannaceae | √ | | | | √ |
| 59. | *Cardiospermum halicacabum* L. | Kapal phodi | Sapindaceae | | √ | √ | | |

| Sl. No. | Botanical Name | Local Name | Family | Exotic | Indi-genous | Wild | Culti-vated | Cultivated and Wild Both |
|---|---|---|---|---|---|---|---|---|
| 60. | *Carica papaya* L. | Papai | Caricaceae | √ | | | √ | |
| 61. | *Cascabella thevetia* (L.) Lippold | Pivla Kanher | Apocynaceae | √ | | | √ | |
| 62. | *Cassia absus* L. | Tarhotana-mama | Caesalpiniaceae | | √ | √ | | |
| 63. | *Cassia auriculata* L | Aawali | Caesalpiniaceae | | √ | √ | | |
| 64. | *Cassia fistula* L. | Bahava; Baswan | Caesalpiniaceae | | √ | √ | | |
| 65. | *Cassia occidentalis* L. | Dukkar sheng; Ran-takla | Caesalpiniaceae | √ | | √ | | |
| 66. | *Cassia sophera* L. | Ran-takla | Caesalpiniaceae | | √ | √ | | |
| 67. | *Cassia tora* L. | Takla; Tarota; Powadya | Caesalpiniaceae | √ | | √ | | |
| 68. | *Catharanthus roseus* (L.) G.Don | Sadafuli | Apocynaceae | √ | | | √ | |
| 69. | *Celastrus paniculata* Willd. | Jyotishmati; Malkangani | Celatraceae | | √ | √ | | |
| 70. | *Celosia argentea* L. | Kurdu; Kirdu | Amranthaceaae | | √ | √ | | |
| 71. | *Centella asiatica* L. | Brahmi. | Apiaceae | | √ | | | √ |
| 72. | *Chlorophytum borivilianum* Sant. and Fernand. | Safed Musali | Liliaceae | | √ | | √ | |
| 73. | *Chrysanthemum indicum* L. | Shevanti | Asteraceae | √ | | | √ | |
| 74. | *Cicer arietinum* L. | Harbhara | Fabaceae | √ | | | √ | |
| 75. | *Cissampelos pareira* L. | Pahadvel; Paathaa | Menispermaceae | | √ | √ | | |
| 76. | *Cissus quadrangularis* L | Kandvel. | Vitaceae | | √ | √ | | |
| 77. | *Citrullus colocynthis* Sch. | Kadu-Indrayan | Cucurbitaceae | | √ | √ | | |
| 78. | *Citrullus lanatus* (Thunb.) Matsura and Nakai | Tarbuj | Cucurbitaceae | √ | | | √ | |

| Sl. No. | Botanical Name | Local Name | Family | Exotic | Indi-genous | Wild | Culti-vated | Cultivated and Wild Both |
|---|---|---|---|---|---|---|---|---|
| 79. | *Citrus aurantifolia* (Christm. and Panz.) Sw. | Limbu | Rutaceae | | √ | | √ | |
| 80. | *Citrus aurantium* L. | Santra | Rutaceae | | √ | | √ | |
| 81. | *Cleome gynandra* L | Pandhri Tilwan | Capparidaceae | | √ | | √ | |
| 82. | *Cleome viscosa* L | Pivli-Tilwan | Capparidaceae | | √ | √ | | |
| 83. | *Clitoria biflora* Dalz. | Garali | Fabaceae | | √ | √ | | |
| 84. | *Clitoria ternatea* L. | Gokarna | Fabaceae | √ | | | √ | |
| 85. | *Coccinia grandis* (L.) Voigt | Tondli, Tongal | Cucurbitaceae | | √ | | | √ |
| 86. | *Cocculus hirsutus* (L.) Theob. | Para-Vel; Vasanvel | Menispermaceae | | √ | √ | | |
| 87. | *Cocos nucifera* L. | Naral | Arecaceae | √ | | | | √ |
| 88. | *Corchorus capsularis* L. | Chuncha. | Tiliaceae | | √ | √ | | |
| 89. | *Cordia dichotoma* Forst. *f.* | Bhokar | Boraginaceae | | √ | √ | | |
| 90. | *Coriandrum sativum* L. | Kothimbir, Dhane | Umbellifereae | | √ | | √ | |
| 91. | *Costus specious* (Koen.) J. E. Sm. | Pev | Zingiberaceae | | √ | √ | | |
| 92. | *Crataeva adansonii* DC. | Waiwarna; Varuna | Capparidaceae | | √ | √ | | |
| 93. | *Crinum asiaticum* L. | Gadhani-Kanda; Gadambi Kanda | Amaryllidaceae | | √ | √ | | |
| 94. | *Cucumis melo* L. | Kharbuj. | Cucurbitaceae | | √ | | √ | |
| 95. | *Cucumis sativus* L. | Kakdi. | Cucurbitaceae | | √ | | √ | |
| 96. | *Cucurbita maxima* Duch. ex Lam. | Bhopla. | Cucurbitaceae | | √ | | √ | |
| 97. | *Curculigo orchioides* Gaertn. | Kala musli; Taalmuli | Hypoxidaceae | | √ | √ | | |
| 98. | *Curcuma amada* Roxb. | Ambe halad. | Zingiberaceae | | √ | | √ | |
| 99. | *Curcuma inodora* Blatt. | Ran halad | Zingiberaceae | | √ | √ | | |

| Sl. No. | Botanical Name | Local Name | Family | Exotic | Indi-genous | Wild | Culti-vated | Cultivated and Wild Both |
|---|---|---|---|---|---|---|---|---|
| 100. | *Curcuma longa* L. | Halad | Zingiberaceae | | √ | | √ | |
| 101. | *Cymbopogon citratus* (DC.) Stapf | Gavti chaha | Poaceae | | √ | | | √ |
| 102. | *Cymbopogon martinii* (Roxb.) Wats. | Rosha gavat | Poaceae | | √ | √ | | |
| 103. | *Cynodon dactylon* (L.) Pers. | Durva. | Poaceae | | √ | √ | | |
| 104. | *Cyperus rotundus* L | Nagarmotha | Cyperaceae | | √ | √ | | |
| 105. | *Dalbergia sissoo* Roxb. ex DC. | Sisam | Fabaceae | | √ | √ | | |
| 106. | *Datura innoxia* Mill. | Pandhara dhotra. | Solanaceae | | √ | √ | | |
| 107. | *Datura metel* L. | Kala dhotra | Solanaceae | √ | | √ | | |
| 108. | *Daucus carota* L. | Gajar | Apiaceae | √ | | | √ | |
| 109. | *Delonix regia* (Boj. ex Hook.) Raf. | Gulmohar | Caesalpiniaceae | √ | | | √ | |
| 110. | *Desmodium gangeticum* (L.) DC. | Salvan; Shaalparni | Fabaceae | | √ | √ | | |
| 111. | *Diospyros melanoxylon* Roxb | Temburni | Ebenaceae | | √ | √ | | |
| 112. | *Diospyros montana* Roxb | Pali | Ebenaceae | | √ | √ | | |
| 113. | *Diplocyclos palmatus* (L.) Jeffrey | Shivalingi; Gurul | Cucurbitaceae | | √ | √ | | |
| 114. | *Dolichandrone falcata* (Wall. ex DC) Seem. | Medshing; Marshing | Bignoniaceae | | √ | √ | | |
| 115. | *Drimia indica* (Roxb.) Jessop | Ran Kanda | Liliaceae | | √ | √ | | |
| 116. | *Echinops echinatus* Roxb. | Utanti | Asteraceae | | √ | √ | | |
| 117. | *Eclipta prostrata* (L.) L. | Maka | Asteraceae | | √ | √ | | |
| 118. | *Elaeocarpus sphaericus* Gaertn. | Rudraksh | Elaeaocarpaceae | | √ | √ | | |

| Sl. No. | Botanical Name | Local Name | Family | Exotic | Indi-genous | Wild | Culti-vated | Cultivated and Wild Both |
|---|---|---|---|---|---|---|---|---|
| 119. | *Eleusine coracana* (L.) Gaertn. | Nachni | Poaceae | | √ | | √ | |
| 120. | *Emblica officinalis* Gaertn. | Avla | Euphorbiaceae | | √ | | √ | |
| 121. | *Enicostema axillare* (Lam.) Raynal | Chota-karait; Nai | Gentianaceae | | √ | √ | | |
| 122. | *Euphorbia hirta* L. | Dudhi | Euphorbiaceae | √ | | √ | | |
| 123. | *Evolvulus alsinoides* (L.) L. | Vishnukrant | Convolvulaceae | | √ | √ | | |
| 124. | *Ficus benghalensis* L. | Vad | Moraceae | | √ | √ | | |
| 125. | *Ficus racemosa* L. | Umbar; Audumbar | Moraceae | | √ | √ | | |
| 126. | *Ficus religiosa* L. | Pimpal | Moraceae | | √ | √ | | |
| 127. | *Ficus virens* Ait. | Pimpri | Moraceae | | √ | √ | | |
| 128. | *Flacourtia indica* (Burm. *f.*) Merr. | Atrun | Flacourtiaceae | | √ | √ | | |
| 129. | *Garnga pinnata* Roxb. | Kakad, Kakod | Burseraceae | | √ | √ | | |
| 130. | *Gloriosa superba* L. | Kal-lawi | Liliaceae | | √ | √ | | |
| 131. | *Gmelina arborea* Roxb. | Shivan | Verbenaceae | | √ | √ | | |
| 132. | *Gossypium arboreum* L. | Devkapas | Malvaceae | | √ | √ | | |
| 133. | *Gossypium herbaceum* L. | Kapas; Kapus | Malvaceae | | √ | | √ | |
| 134. | *Haldina cordifolia* (Roxb.) Ridsd. | Haldu | Rubiaceae | | √ | √ | | |
| 135. | *Hardwickia binata* Roxb. | Anjan | Caesalpiniaceae | | √ | √ | | |
| 136. | *Helicteres isora* L. | Murudsheng | Sterculiaceae | | √ | √ | | |
| 137. | *Hemidesmus indicus* (L.) Schult. | Sariva; Anantamul | Asclepiadaceae | | √ | √ | | |
| 138. | *Hibiscus rosa-sinensis* L. | Jasvand | Malvaceae | √ | | | √ | |

| Sl. No. | Botanical Name | Local Name | Family | Exotic | Indi-genous | Wild | Culti-vated | Cultivated and Wild Both |
|---|---|---|---|---|---|---|---|---|
| 139. | *Hibiscus sabdariffa* L. | Lal ambadi | Malvaceae | | √ | √ | | |
| 140. | *Holarrhena pubescens* (Buch.-Ham.) Wall. ex G. Don | Dahikudi | Apocynaceae | | √ | √ | | |
| 141. | *Holoptelea integrifolia* Planch. | Papda | Ulmaceae | | √ | √ | | |
| 142. | *Hymenodictyon ovixense* (Roxb.) Mabberly | Madala | Rubiaceae | | √ | √ | | |
| 143. | *Ipomoea illustris* (Clarke) Prain. | Kumrhoney | Convolvulaceae | | | √ | | |
| 144. | *Ipomoea nil* (L.) Roth | Nil-pushpa | Convolvulaceae | | √ | √ | | |
| 145. | *Ixora coccinea* L. | Bokadi | Rubiaceae | √ | | √ | | |
| 146. | *Jasminum officinale* L. | Chameli. | Oleaceae | | √ | | √ | |
| 147. | *Jasminum sambac* (L.) Ait. | Mogra | Oleaceae | | √ | | √ | |
| 148. | *Jatropha curcas* L. | Moghli errand | Euphorbiaceae | √ | | | √ | |
| 149. | *Justicia adhatoda* L. | Adulasa | Acanthaceae | | √ | | | √ |
| 150. | *Kalanchoe pinnata* (Lam.) Pers. | Panphuti | Crassulaceae | | √ | | √ | |
| 151. | *Kohautia aspera* (Heyne ex Roth) Brem. | Khetpara | Rubiaceae | √ | | √ | | |
| 152. | *Lagenaria siceraria* (Mol.) Standl. | Dudhi | Cucurbitaceae | | √ | | √ | |
| 153. | *Lagerstroemia microcarpa* Wight | Nana Bondara; Hina | Lythraceae | | √ | √ | | |
| 154. | *Lantana camara* L. | Ghaneri, Gansubai. | Verbenaceae | √ | | √ | | |
| 155. | *Lawsonia inermis* L. | Kadu mendi; Mehandi | Lythraceae | √ | | √ | | |

| Sl. No. | Botanical Name | Local Name | Family | Exotic | Indi-genous | Wild | Culti-vated | Cultivated and Wild Both |
|---|---|---|---|---|---|---|---|---|
| 156. | *Leucas cephalotes* (Roth) Spr. | Ghumma. | Lamiaceae | | √ | √ | | |
| 157. | *Limonia acidissima* L. | Kavath | Rutaceae | | √ | √ | | |
| 158. | *Luffa acutangula* (L.) Roxb. | Shirale | Cucurbitaceae | | √ | √ | | |
| 159. | *Madhuca longifolia* (Koen.) Mac Bride | Moha, Mahu, Mahuda | Sapotaceae | | √ | √ | | |
| 160. | *Mallotus philippensis* (Lam.) Muell-Arg. | Shendri | Euphorbiaceae | | √ | √ | | |
| 161. | *Mangifera indica* L. | Amba. | Anacardiaceae | | √ | | | √ |
| 162. | *Martynia annua* L. | Winchu. | Martyniaceae | √ | | √ | | |
| 163. | *Melia azedarach* L. | Bakana neem | Meliaceae | √ | | √ | | |
| 164. | *Melilotus alba* Medik. ex Desr. | Ran methi | Fabaceae | | √ | √ | | |
| 165. | *Merremia dissecta* (Jacq.) Hallier *f.* | | Convolvulaceae | | √ | √ | | |
| 166. | *Michelia champaca* L. | Sonchapha | Magnoliaceae | | √ | | √ | |
| 167. | *Mimosa hamata* Willed. | Arati | Mimosaceae | | √ | √ | | |
| 168. | *Mimusops elengi* L. | Bakul. | Sapotaceae | | √ | √ | | |
| 169. | *Mirabilis jalapa* L. | Gulbakshi | Nyctaginaceae | √ | | | √ | |
| 170. | *Mitragyna parvifolia* (Roxb.) Korth. | | Rubiaceae | | √ | √ | | |
| 171. | *Momordica dioica* Roxb. ex Willd. | Kartoli; Katurle | Cucurbitaceae | | √ | √ | | |
| 172. | *Moringa oleifera* Lam. | Shevga | Moringaceae | | √ | | √ | |
| 173. | *Mucuna pruriens* (L.) DC. | Kahj-kuili | Fabaceae | | √ | √ | | |
| 174. | *Mukia maderaspatana* L. | Ankhphod | Cucurbitaceae | | √ | √ | | |

| Sl. No. | Botanical Name | Local Name | Family | Exotic | Indi-genous | Wild | Culti-vated | Cultivated and Wild Both |
|---|---|---|---|---|---|---|---|---|
| 175. | *Musa paradisiaca* L. | Kel | Musaceae | | √ | | √ | |
| 176. | *Nelumbo nucifera* Gaertn. | Kamal | Nelumbonaceae | | √ | | | √ |
| 177. | *Nerium indicum* Mill. | Kanher | Apocynaceae | √ | | | √ | |
| 178. | *Nyctanthes arbortristis* L. | Parijatak | Oleaceae | | √ | | √ | |
| 179. | *Nymphaea nouchali* Burm.*f.* | Kamal; Kamay | Nymphaeaceae | | √ | | | √ |
| 180. | *Nymphaea pubescens* Willd | Kamal | Nymphaeaceae | | √ | | | √ |
| 181. | *Ocimum basilicum* L. | Sabja | Lamiaceae | | √ | √ | | |
| 182. | *Ocimum gratissimum* L. | Ram Tulasi | Lamiaceae | | √ | √ | | |
| 183. | *Ocimum tenuiflorum L.* | Krishnatulas | Lamiaceae | | √ | √ | | |
| 184. | *Opuntia elatior* Mill. | Nivdung | Cactaceae | √ | | √ | | |
| 185. | *Oroxylum indicum* (L.) Vent. | Tetu. | Bignoniaceae | | √ | √ | | |
| 186. | *Oryza sativa* L. | Tandul, Dhan | Poaceae | | √ | | √ | |
| 187. | *Oxalis corniculata* L. | Ambuti | Oxalidaceae | | √ | √ | | |
| 188. | *Pandanus odoratissimus* L. *f.* | Kevada | Pandanaceae | | √ | √ | | |
| 189. | *Passiflora edulis* Sims | Krishna Kamal | Passifloraceae | √ | | | √ | |
| 190. | *Pedalium murex* L. | Mothi-Gokhru, Ubha Gokharu | Pedaliaceae | √ | | √ | | |
| 191. | *Pennisetum americanum* L. | Bajaras, Bajari | Poaceae | √ | | | √ | |
| 192. | *Pergularia daemia* (Forssk.) Chiov. | Utaran; Rota | Asclepiadoideae | | √ | √ | | |
| 193. | *Phoenix dactylifera* (L) Roxb. | Khajur | Arecaceae | | √ | | √ | |
| 194. | *Phyla nodiflora* (L.) Greene | Mundi | Verbenaceae | | √ | √ | | |
| 195. | *Phyllanthus acidus* (L.) K. Skeels | Rai Awala | Euphorbiaceae | √ | | | √ | |
| 196. | *Phyllanthus amarus* L. | Bhui Awla | Euphorbiaceae | | √ | | √ | |

| Sl. No. | Botanical Name | Local Name | Family | Exotic | Indi-genous | Wild | Culti-vated | Cultivated and Wild Both |
|---|---|---|---|---|---|---|---|---|
| 197. | *Physalis micrantha* L. | Koyi | Solanaceae | | √ | √ | | |
| 198. | *Piper betle* L. | Paan | Piperaceae | | √ | | √ | |
| 199. | *Plumbago zeylanica* L. | Chitrak | Plumbaginaceae | | √ | √ | | |
| 200. | *Plumeria rubra* L. forma *acuminata* (Ait.) Santapaul and Irani ex Shah | Pandhara Chapha | Apocynaceae | √ | | | √ | |
| 201. | *Polyalthia longifolia* Sonn. | Ashok | Annonaceae | | √ | | √ | |
| 202. | *Pongamia pinnata* (L.) Pierre | Karanj | Lamiaceae | | √ | √ | | |
| 203. | *Prosopis cineraria* (L.) Druce | Saundad, Samadi | Mimosaceae | | √ | √ | | |
| 204. | *Psidium guajava* L. | Peru | Myrtaceae | √ | | | √ | |
| 205. | *Pterocarpus marsupium* Roxb. | Bibla; Bijjaka | Fabaceae | | √ | √ | | |
| 206. | *Pueraria tuberosa* (Roxb. ex Willd.) DC. | Bhui Kohala | Fabaceae | | √ | √ | | |
| 207. | *Punica granatum* L. | Dalimb | Punicaceae | √ | | | √ | |
| 208. | *Quisqualis indica* L. | Madhumalti | Combretaceae | √ | | | √ | |
| 209. | *Raphanus sativus* L. | Mula | Brassicaceae | √ | | | √ | |
| 210. | *Ricinus communis* L. | Erand | Euphorbiaceae | √ | | √ | | √ |
| 211. | *Rosa damascena* Mill. | Gulab | Rosaceae | √ | | | √ | |
| 212. | *Saccharum officinarum* L. | Uss | Poaceae | | √ | | √ | |
| 213. | *Salvadora persica* L. | Pilu | Salvadoraceae | | √ | √ | | |
| 214. | *Santalum album* L. | Chandan | Santalaceae | | √ | √ | | |
| 215. | *Sapindus emarginatus* Vahl. | Ritha | Sapindaceae | | √ | √ | | |
| 216. | *Saraca asoca* (Roxb.) de Wilde | Samudra Ashok | Caesalpiniaceae | | √ | √ | | |

| Sl. No. | Botanical Name | Local Name | Family | Exotic | Indi-genous | Wild | Culti-vated | Cultivated and Wild Both |
|---|---|---|---|---|---|---|---|---|
| 217. | *Schleichera oleosa* (Lour.) Oken | Kusum | Sapindaceae | | √ | √ | | |
| 218. | *Securinega virosa* (Roxb. ex Willd.) Baill. | Pandharphali | Euphorbiaceae | | √ | √ | | |
| 219. | *Semecarpus anacardium* L. | Bhilava | Anacardiaceae | | √ | √ | | |
| 220. | *Sesamum orientale* L. | Til | Pedaliaceae | | √ | | √ | |
| 221. | *Sesbania grandiflora* (L.) Poir. | Hadga | Fabaceae | √ | | √ | | |
| 222. | *Sesbania sesban* (L.) Merr. | Sheuri, Shewari | Fabaceae | | √ | √ | | |
| 223. | *Setaria italica* (L.) P. Beauv. | Bhandi | Poaceae | | √ | √ | | |
| 224. | *Sida acuta* Burm. *f.* | Bokya | Malvaceae | | √ | √ | | |
| 225. | *Sida cordata* (Burm.*f.*) Borss. | Bala | Malvaceae | | √ | √ | | |
| 226. | *Sida spinosa* L. | Gangeti; Kantalobata | Malvaceae | | √ | √ | | |
| 227. | *Solanum anguivi* Lam. | Mothi ringni | Solanaceae | | √ | √ | | |
| 228. | *Solanum melongena* L. | Vangi | Solanaceae | | √ | | √ | |
| 229. | *Solanum nigrum* L. | Kangani | Solanaceae | | √ | √ | | |
| 230. | *Solanum virginianum* L. | Bhuiringani. | Solanaceae | | √ | √ | | |
| 231. | *Sorghum vulgare* Pers. | Jowar. | Poaceae | | √ | | √ | |
| 232. | *Spilanthus calva* DC. | Akkalkadha | Asteraceae | | √ | √ | | |
| 233. | *Spondias pinnata* (L. *f.*) Kurz. | Ambada | Anacardiaceae | | √ | √ | | |
| 234. | *Sterculia urens* Roxb. | Kadai; Kadoni; Kevda; Kudal | Sterculiaceae | | √ | √ | | |
| 235. | *Syzygium cumini* (L.) Skeels | Jambhul. | Myrtaceae | | √ | | | √ |
| 236. | *Tabernaemontana divaricata* (L.) R. Br. | Tagar; Chandani | Apocynaceae | | √ | | √ | |

| *Sl. No.* | *Botanical Name* | *Local Name* | *Family* | *Exotic* | *Indi-genous* | *Wild* | *Culti-vated* | *Cultivated and Wild Both* |
|---|---|---|---|---|---|---|---|---|
| 237. | *Tagetes erecta* L. | Zendu | Asteraceae | | √ | | √ | |
| 238. | *Tamarindus indica* L. | Imli | Caesalpiniaceae | | √ | √ | | |
| 239. | *Tectona grandis* L. | Sag | Verbenaceae | | √ | | | √ |
| 240. | *Tephrosia purpurea* Pers. | Sarpankha | Fabaceae | | √ | √ | | |
| 241. | *Terminalia arjuna* (Roxb. ex DC.) Wight and Arn. | Arjun | Combretaceae | | √ | √ | | |
| 242. | *Terminalia bellirica* (Gaertn.) Roxb. | Beheda | Combretaceae | | √ | √ | | |
| 243. | *Thespesia populnea* (L.) Soland. ex Corr. | Paras pimpal | Malvaceae | | √ | √ | | |
| 244. | *Thuja plicata* ex D. Don | Vidya | Cupressaceae | | √ | | √ | |
| 245. | *Tinospora cordifolia* (Willd.) Miers ex Hook. *f.* and Thoms | Gudvel | Menispermaceae | | √ | √ | | |
| 246. | *Tribulus terrestris* L. | Gokhru; Gokshura mool | Zygophyllaceae | | √ | √ | | |
| 247. | *Tricholepis indica* (Willd.) Almeida | Dahan | Asteraceae | | √ | √ | | |
| 248. | *Trichosanthes cucumerina* L. | Jangli padval | Cucurbitaceae | | √ | √ | | |
| 249. | *Trichosanthes tricuspidata* Lour. | Mukul | Cucurbitaceae | | √ | √ | | |
| 250. | *Tricticum aestivum* L. | Gahu | Poaceae | √ | | | √ | |
| 251. | *Tridax procumbens* L. | Dagdi pala | Asteraceae | | √ | √ | | |
| 252. | *Vigna mungo* (L.) Hepper | Udid | Fabaceae | | √ | | √ | |
| 253. | *Vigna radiata* (L.) R. Wilczek | Sona mug | Fabaceae | | √ | | √ | |
| 254. | *Vigna trilobata* (L.) Verdc. | Mung | Fabaceae | | √ | | √ | |
| 255. | *Vitex negundo* L. | Nirgundi | Verbenaceae | | √ | √ | | |

| Sl. No. | Botanical Name | Local Name | Family | Exotic | Indi-genous | Wild | Culti-vated | Cultivated and Wild Both |
|---|---|---|---|---|---|---|---|---|
| 256. | *Vitis vinifera* L. | Draksha | Vitaceae | | √ | | √ | |
| 257. | *Withania somnifera* (L.) Dunal | Ashwagandha | Solanaceae | | √ | √ | | |
| 258. | *Woodfordia fruticosa* (L.) Kurz. | Dhoneyti | Lythraceae | | √ | √ | | |
| 259. | *Wrightia tinctoria* R.Br. | Pandhra kuda | Apocynaceae | | √ | √ | | |
| 260. | *Xanthium indicum* Koen. | Sankeshwar | Asteraceae | | √ | √ | | |
| 261. | *Zea mays* L. | Maka, Makki, Makai | Poaceae | √ | | | √ | |
| 262. | *Zingiber officinale* Rosc. | Adrak | Zingibaraceae | | √ | | √ | |
| 263. | *Zizyphus mauritiana* Lam. | Bor | Rhamnaceae | | √ | | | √ |

of cultivated 103 species out of 264 total species naturally help conserve wild species of the region. Exhaustive pressure is thereby minimized on the wild or natural flora of the region.

## Status Regarding Wild Species

As mentioned earlier, 103 species are being cultivated for various miscellaneous purposes. The rest others (161 species) are wild ones (Table 2). Amongst these wild ones, there are some species which are planted in the region. It means they belong to both catagories as wild and cultigens *e.g. Abrus precatorius* L., *Aegle marmelos* (L.) Corr., *Annona squamosa* L., *Azadirachta indica* A. Juss., *Buchanania lanzan* Spreng., *Clitoria ternatea* L., *Ficus religiosa* L., *Madhuca longifolia* (Koen.) MacBr., *Pongamia pinnata* (L.) Pierre., *Tamarindus indica* Linn., *Tectona grandis* L. *f*. Thus total 149 species are exclusively wild and integral part of the local biodiversity. This trend indicates that indigenous floral elements have been, by trial and errors, brought under cultivation. Thus useful species are made available near human habitations and appropriated also for various religious and sacred purposes.

## Status Regarding Exotic Species

Exotic species are those which do not occur naturally in a region. Thus they are aliens and not indigenous. They invade a particular region unintentionally and are wild or naturalized as such in the region of invasion. Some other exotic species are brought intentionally in a region. They are used for various purposes. They are introduced and domesticated in past. These are part of human activity for their welfare. They supplement the local bioresources or even add sources of income. As many as 119 exotic species are exotic in the area of present study. Majority of them are cultigens, whereas only few species run wild and have become a part of wild flora. All these belong to 14 different parts of the world as noted in the Table 3. Interestingly, their major share is hailed from various parts of American continent (New World). Different regions of the old world are represented by just one or two species each. Exotic species are thus indicative of human activities and contacts. The people of India have developed contacts with other countries or geographical regions outside Indian territory since ancient times. It is definitely astonishing that Indians have had contacts with America, although America is said to be of rather recent discovery (year 1492 AD.) by Columbus. Very few species (09 species) *viz., Amaranthus spinosus* L., *Argemone mexicana* L., *Cassia occidentalis* L., *Cassia tora* L., *Datura metel* L., *Euphorbia hirta* L., *Lantana camara* L., *Martynia annua* L. And *Pedalium murex* L. This fact suggests that some species of American origin are brought intentionally for different material uses in India. This is not the case only. They have long been appropriated for religious purposes and conceived sacred for gods and goddesses in India. For example, (i). *Annona squamosa* L. (Annonaceae): Its fruit are depicted in the hands of goddess Durga, a temple at Aihole (Karnataka) from the 10$^{th}$ century CE (Johannessen and Siming,

**Table 3: Exotic Plants**

| Sl.No. | Botanical Name | Family |
|---|---|---|
| 1. | *Abelmoschus esculentus* (L.) Moench. | Malvaceae |
| 2. | *Achyranthes aspera* L. | Amarantaceae |
| 3. | *Adansonia digitata* L. | Bombacaceae |
| 4. | *Albizia lebbeck* (L.) Willd. | Mimosaceae |
| 5. | *Allium cepa* L. | Liliaceae |
| 6. | *Allium sativum* L. | Liliaceae |
| 7. | *Aloe vera* L. | Liliaceae |
| 8. | *Amaranthus spinosus* L. | Amaranthaceae |
| 9. | *Anethum graveolens* L. | Labiatae |
| 10. | *Annona reticulata* L. | Umbellifereae |
| 11. | *Annona squamosa* L. | Annonaceae |
| 12. | *Arachis hypogaea* L. | Fabaceae |
| 13. | *Argemone mexicana* L. | Papaveranceae |
| 14. | *Barleria prionitis* L. | Acanthaceae |
| 15. | *Bauhinia tomentosa* L. | Caesalpiniaceae |
| 16. | *Bauhinia variegata* L. | Caesalpiniaceae |
| 17. | *Benincasa hispida* (Thunb.) Cogn | Cucurbitaceae |
| 18. | *Boerhavia repens* L. var. *diffusa* (L.) Hook *f.* | Nyctaginaceae |
| 19. | *Bombax ceiba* L. | Bombacaceae |
| 20. | *Borassus flabellifer* L. | Aracaceae |
| 21. | *Bougainvillea spectabilis* Willd. | Nyctaginaceae |
| 22. | *Caesalpinia bonduc* (L.) Roxb. | Caesalpiniaceae |
| 23. | *Caesalpinia pulcherrima* (L.) Sw. | Caesalpiniaceae |
| 24. | *Calotropis gigantea* (L.) R. Br | Asclepiadaceae |
| 25. | *Calotropis procera* (Ait.) R. Br. | Asclepiadaceae |
| 26. | *Canna indica* L. | Cannaceae |
| 27. | *Cardiospermum halicacabum* L. | Sapindaceae |
| 28. | *Carica papaya* L. | Caricaceae |
| 29. | *Cascabella thevetia* (L.) Lippold | Apocynaceae |
| 30. | *Cassia fistula* L. | Caesalpiniaceae |
| 31. | *Cassia occidentalis* L. | Caesalpiniaceae |
| 32. | *Cassia sophera* L. | Caesalpiniaceae |
| 33. | *Cassia tora* L. | Caesalpiniaceae |
| 34. | *Catharanthus roseus* (L.) G.Don | Apocynaceae |
| 35. | *Celosia argentea* L. | Amranthaceaae |
| 36. | *Chrysanthemum indicum* L. | Asteraceae |
| 37. | *Cicer arietinum* L. | Fabaceae |

| Sl.No. | Botanical Name | Family |
|---|---|---|
| 38. | *Cissampelos pareira* L. | Menispermaceae |
| 39. | *Citrullus lanatus* (Thunb.) Matsura and Nakai | Cucurbitaceae |
| 40. | *Citrus aurantifolia* (Christm. and Panz.) Sw. | Rutaceae |
| 41. | *Citrus aurantium* L. | Rutaceae |
| 42. | *Cleome gynandra* L | Capparidaceae |
| 43. | *Cleome viscosa* L | Capparidaceae |
| 44. | *Clitoria ternatea* L. | Fabaceae |
| 45. | *Coccinia grandis* (L.) Voigt | Cucurbitaceae |
| 46. | *Corchorus capsularis* L. | Tiliaceae |
| 47. | *Coriandrum sativum* L. | Umbellifereae |
| 48. | *Cymbopogon citratus* (DC.) Stapf | Poaceae |
| 49. | *Cymbopogon martinii* (Roxb.) Wats. | Poaceae |
| 50. | *Cynodon dactylon* (L.) Pers. | Poaceae |
| 51. | *Cyperus rotundus* L | Cyperaceae |
| 52. | *Datura innoxia* Mill. | Solanaceae |
| 53. | *Datura metel* L. | Solanaceae |
| 54. | *Daucus carota* L. | Apiaceae |
| 55. | *Delonix regia* (Boj. ex Hook.) Raf. | Caesalpiniaceae |
| 56. | *Echinops echinatus* Roxb. | Asteraceae |
| 57. | *Eclipta prostrata* (L.) L. | Asteraceae |
| 58. | *Eleusine coracana* (L.) Gaertn. | Poaceae |
| 59. | *Euphorbia hirta* L. | Euphorbiaceae |
| 60. | *Evolvulus alsinoides* (L.) L. | Convolvulaceae |
| 61. | *Gmelina arborea* Roxb. | Verbenaceae |
| 62. | *Gossypium herbaceum* L. | Malvaceae |
| 63. | *Hibiscus rosa-sinensis* L. | Malvaceae |
| 64. | *Hibiscus sabdariffa* L. | Malvaceae |
| 65. | *Ipomoea nil* (L.) Roth | Convolvulaceae |
| 66. | *Jasminum officinale* L. | Oleaceae |
| 67. | *Jasminum sambac* (L.) Ait. | Oleaceae |
| 68. | *Jatropha curcas* L. | Euphorbiaceae |
| 69. | *Kalanchoe pinnata* (Lam.) Pers. | Crassulaceae |
| 70. | *Lagenaria siceraria* (Mol.) Standl. | Cucurbitaceae |
| 71. | *Lantana camara* L. | Verbenaceae |
| 72. | *Leucas cephalotes* (Roth) Spr. | Lamiaceae |
| 73. | *Luffa acutangula* (L.) Roxb. | Cucurbitaceae |
| 74. | *Martynia annua* L. | Martyniaceae |
| 75. | *Melia azedarach* L. | Meliaceae |
| 76. | *Melilotus alba* Medik. ex Desr. | Fabaceae |

| Sl.No. | Botanical Name | Family |
|---|---|---|
| 77. | *Merremia dissecta* (Jacq.) Hallier *f.* | Convolvulaceae |
| 78. | *Mimosa hamata* Willed. | Mimosaceae |
| 79. | *Mirabilis jalapa* L. | Nyctaginaceae |
| 80. | *Moringa oleifera* Lam. | Moringaceae |
| 81. | *Mucuna pruriens* (L.) DC. | Fabaceae |
| 82. | *Nerium indicum* Mill. | Apocynaceae |
| 83. | *Opuntia elatior* Mill. | Cactaceae |
| 84. | *Oxalis corniculata* L. | Oxalidaceae |
| 85. | *Passiflora edulis* Sims. | Passifloraceae |
| 86. | *Pedalium murex* L. | Pedaliaceae |
| 87. | *Phoenix dactylifera* (L) Roxb. | Arecaceae |
| 88. | *Phyla nodiflora* (L.) Greene | Verbenaceae |
| 89. | *Phyllanthus acidus* (L.) K. Skeels | Euphorbiaceae |
| 90. | *Piper betle* L. | Piperaceae |
| 91. | *Plumbago zeylanica* L. | Plumbaginaceae |
| 92. | *Plumeria rubra* L. forma *acuminata* (Ait.) Santapaul and Irani ex Shah | Apocynaceae |
| 93. | *Psidium guajava* L. | Myrtaceae |
| 94. | *Punica granatum* L. | Punicaceae |
| 95. | *Quisqualis indica* L. | Combretaceae |
| 96. | *Raphanus sativus* L. | Brassicaceae |
| 97. | *Ricinus communis* L. | Euphorbiaceae |
| 98. | *Rosa damascena* Mill. | Rosaceae |
| 99. | *Sesbania grandiflora* (L.) Poir. | Fabaceae |
| 100. | *Sesbania sesban* (L.) Merr. | Fabaceae |
| 101. | *Setaria italica* (L.) P. Beauv. | Poaceae |
| 102. | *Sida acuta* Burm. *f.* | Malvaceae |
| 103. | *Sida cordata* (Burm.*f.*) Borss. | Malvaceae |
| 104. | *Sida spinosa* L. | Malvaceae |
| 105. | *Solanum melongena* L. | Solanaceae |
| 106. | *Solanum nigrum* L. | Solanaceae |
| 107. | *Solanum virginianum* L. | Solanaceae |
| 108. | *Sorghum vulgare* Pers. | Poaceae |
| 109. | *Spondias pinnata* (L. *f.*) Kurz. | Anacardiaceae |
| 110. | *Tagetes erecta* L. | Asteraceae |
| 111. | *Tamarindus indica* L. | Caesalpiniaceae |
| 112. | *Tribulus terrestris* L. | Zygophyllaceae |
| 113. | *Tricticum aestivum* L. | Poaceae |
| 114. | *Tridax procumbens* L. | Asteraceae |

| Sl.No. | Botanical Name | Family |
|---|---|---|
| 115. | *Vitex negundo* L. | Verbenaceae |
| 116. | *Vitis vinifera* L. | Vitaceae |
| 117. | *Xanthium indicum* Koen. | Asteraceae |
| 118. | *Zea mays* L. | Poaceae |
| 119. | *Zizyphus mauritiana* Lam. | Rhamnaceae |

1998). This is also corroborated by (a) its fruits depicted in the hands of Lord Vishnu from Bengal (b), Kartikaya in Madhukesvara temple (Mukhalingam), Karnataka, Andhra Pradesh (8$^{th}$ century CE.), (c) Lord Shiva in Kakatiya temple (12$^{th}$ century CE.) and Kubera at the Holyasaleshvara temple (12$^{th}$ century CE.) from Karnataka (Nandita Krishna and Amirthalingam, 2014). (ii). *Zea mays* L. (Poaceae): Its ears are depicted at Hoysala temple (near Mysore), Karnataka built in 12$^{th}$ and 13$^{th}$ century AD (Johannessen and Parker, 1989). They also depicted in left hand of 8-armed dancing Vishnu in his female form 'Mohini' at Laxmi Narsimha temple, Nugghalli (Karnataka) 12$^{th}$-13$^{th}$ century AD (Gupta, 1996). (iii) *Datura metel* L. (Solanaceae): Its flower is depicted on a headdress of Lord Shiva (Geeta and Waleed, 2007). Some exotic species have been recovered from archaeological sites indicating their introduction and domestication in India *e.g.* (i) *Ricinus communis* L. (Euphorbiaceae): It is recorded earliest between 1800 and 1300 BC. in Harappan site of Hulas (Saraswat, 1992). (ii) *Cicer arietinum* L. (Fabaceae) : It was introduced as early as 2000 BC (Vishnu-Mittre, 1974). (iv) *Allium cepa* L. (Liliaceae): Archaeological remains of onion seeds are recorded in Middle Gagetic Plains Wain, Ballia and Raja-Nala-Ka-Tila in Uttar Pradesh during 800-1600 BC (Saraswat, 2005; Singh and Nigam, 2017). (v) *Allium sativum* L. (Liliaceae): It was introduced in India during Mohenjo-Daro period about 300 BC (Singh and Nigam, 2017). All these empirical evidances are suggestive of introduction of exotic plants long back even before Columbian period and indicate their antiquity. Discussion on other exotic species on this line is avoided for the take of brevity of presentation. This area of research can be extended as an exclusive topic of investigation.

## Plant Species Useful for Medicine

The investigators made enquiry about medicinal uses to the priest, devotees, some people residing in temple yards, some tribal pujara (worshipper), *etc.* Nearly 708 use-reports are recorded for medicinal purposes. They informed a total 232 species useful to combat various diseases as noted under enumeration. Usually, they informed medicinal utility of various plant parts *viz.* stems (65 species), roots (103 species), leaves (116 species), fruits (79 species), flowers (43 species) and seeds (53 species). Leaves are employed on large scale, whereas other parts are employed in descending order as roots, fruits, stem, seeds and flowers. Out of 232 species medicinally useful, 68 species are trees. Trees are rather perennial sources and are employed throughout a year. Herbaceous and shrubby floral elements predominate local plant diversity but they are useful in different months of the year. The plant taxa which are usually planted in temple yards have easy access. They thus sustain the needy people to combat human afflictions in reasonable time (Table 5).

## Plant Species Useful for Sacred Purposes

The present inventory related as many as 163 species useful for sacred and religious purposes (Table 4). Different parts form offerings to gods and goddess and local deities, called gramdevata (village deities). Village deities are worshipped more especially by rural and tribal people as noted in Appendices: I, II, III and VII. Some form of worships are also observed during festivals, pilgrims and yatra (Chariot fair of gods), besides religion-wise or caste-wise rites, rituals, customs and social ceremonies like birth and marriages. Fairs are observed on some particular days in the name of gods or goddess. In such cases, gods and goddesses are worshiped together by the entire village community. Individual worships are also noted by some families on certain occasions. Some articles of metals, some figures, recipes or plant part are used as symbolism during various worships (Appendix VII). Various plants that constitute divine offerings are : roots (09 species), stems(13 species), seeds (21 species), leaves (33 species), flowers (44 species) and fruits (53 species). Nearly 475 use-reports are recorded for sacred and socio-religious purposes. On some worships, plant parts are not offered as such but the individual plant species is worshipped ceremonially *e.g.* (i) *Ficus benghalensis* L., on Vat-Savitri Pournima by Hindus, (ii) *Ocimum tenuiflorum* L. by Hindus, *etc.* Plants thus play a notable role in socio-religious ceremonies and worships. This has been observed almost in every society. Plant parts in the form of offerings by the Indians are also recorded from the Vedic period. Several instances and references to the uses of plants are noted in the time-honored Indian epics (*e.g.* Ramayana, Mahabharata) and other Sanskrit scriptures (*e.g.* Kautilya's Arthashashra, Manusmruti) and various puranic literature (*e.g.* Vaman Purana, Shiv Purana, Matsya Purana, *etc.*). It is generally believed that the plants used in different rituals are endowed with peculiar virtues and hence they are venerated. Several flowers, fruits, leaves or entire plants have been used as incense, ceremonial sacrifices, votive-offerings to deities for warding off evil spirits and for providing fertility and boon of a mate child, good luck, prosperity and protection (Patil, 2020).

## Plant Species Useful for Miscellaneous Purposes

This inventory also recorded the various species under consideration from the point of ethnobotanical and economic significance which are generally helpful for sustenance locally. Total 44 species are reported useful for miscellaneous purposes as stated in the text. Various plant parts are used by the local people. Such useful parts are: leaves (20 species), fruits (19 species), stems (17 species), seeds (11species), roots and flowers (10 species each), bark (03 species) and

**Table 4: Plant Parts Used for Sacred Purposes**

| Sl.No. | Botanical Name | Local Name | Fruits | Leaves | Flowers | Stem | Roots | Seeds |
|---|---|---|---|---|---|---|---|---|
| 1. | *Abelmoschus esculentus* (L.) Moench. | Bhendi | √ | | | | | |
| 2. | *Abelmoschus manihot* (L.) Medik. | Kasturi bhendi, Ran bhendi | | | | | | √ |
| 3. | *Abitulon indicum* (L.) | Sikka, Atti. | √ | | | | | |
| 4. | *Abrus precatorius* L. | Gunj, Ratti | | | | | | √ |
| 5. | *Acacia chundra* (Roxb. ex Rottle.) Willd. | Khair | | | | √ | | √ |
| 6. | *Acacia ferruginea* DC. | Devbhabul | | | | √ | | |
| 7. | *Acacia nilotica* (L.) Del. | Ramkathi. | | | | √ | | |
| 8. | *Acalypha indica* L. | Kokali | | | | | | |
| 9. | *Achyranthes aspera* L. | Aghada, Chirchir | | √ | | | | |
| 10. | *Acorus calamus* L. | Vekhand; Yekhand | | | | √ | | |
| 11. | *Adansonia digitata* L. | Gorakh chinch | √ | | | | | |
| 12. | *Aegle marmelos* (L.) Corr. | Bel phal, Bel mool, Bel. | √ | √ | | | | |
| 13. | *Ailanthus excelsa* Roxb. | Maharuk | | | | √ | | |
| 14. | *Albizia lebbeck* (L.) Willd. | Shirish | | | | | | |
| 15. | *Allium cepa* L. | Kanda | | | | | √ | |
| 16. | *Allium sativum* L. | Lasun | | | | | √ | |
| 17. | *Aloe vera* L. | Korphad | | √ | | | | |
| 18. | *Alstonia scholaris* R.Br. | Saptaparn | | √ | | | | |
| 19. | *Amaranthus hybridus* L | Rajgira | | | | | | √ |
| 20. | *Amaranthus spinosus* L. | Kate – math | | √ | | | | |
| 21. | *Ammannia baccifera* L. | Aagia | | | | | | |
| 22. | *Andrographis echioides* (L.) Nees. | Kalmegh | | | | | | |

| Sl.No. | Botanical Name | Local Name | Fruits | Leaves | Flowers | Stem | Roots | Seeds |
|---|---|---|---|---|---|---|---|---|
| 23. | *Anethum graveolens* L. | Balant shepu | | √ | | | | |
| 24. | *Anisomeles malabarica* R. Br. | Kalatumba | | | | | | |
| 25. | *Annona reticulata* L. | Ramphal. | √ | | | | | |
| 26. | *Annona squamosa* L. | Sitaphal | √ | | | | | |
| 27. | *Anogeissus latifolia* Wall. | Dhavda | | | | | | |
| 28. | *Apium graveolens* L. | Shepu | | | | | | √ |
| 29. | *Arachis hypogaea* L. | Bhuimug | | | | | | √ |
| 30. | *Argemone mexicana* L. | Pivla Dhotra | √ | | | | | |
| 31. | *Argyreia nervosa* (Burm. *f.*) Bojer | Samudrashok | | | | | | |
| 32. | *Aristolochia bracteolata* Lam. | Kidamar | | | | | | |
| 33. | *Asparagus racemosus* Willd | Shatavari, Chinyar | | | | | | |
| 34. | *Azadirachta indica* A. Juss. | Kadu neem | | √ | | √ | | |
| 35. | *Bacopa monnieri* Penn. | Brahmi | | | | | | |
| 36. | *Balanitis aegyptiaca* (L.) Delile | Hinganbet | | | | | | |
| 37. | *Baliospermum raziana* Keshav *et.* Yog | Danti | | | | | | |
| 38. | *Bambusa arundinacea* (Retz.)Willd. | Kalak; Bambu. | | | | √ | | |
| 39. | *Barleria lawii* L. | Akhar | | | | | | |
| 40. | *Barleria prionitis* L. | Kate Koranti. | | | √ | | | |
| 41. | *Bauhinia racemosa* Lamk. | Apta | | √ | | | | |
| 42. | *Bauhinia tomentosa* L. | Ashundar | | | | | | |
| 43. | *Bauhinia variegata* L. | Kanchan. | | | √ | | | |
| 44. | *Benincasa hispida* (Thunb.) Cogn | Kohla. | √ | | | | | |
| 45. | *Boerhavia repens* L. var. *diffusa* (L.) Hook *f.* | Punarnawa | | | | | | |

| Sl.No. | Botanical Name | Local Name | Fruits | Leaves | Flowers | Stem | Roots | Seeds |
|---|---|---|---|---|---|---|---|---|
| 46. | *Bombax ceiba* L. | Kate Sawar, Hawar | | | | | | |
| 47. | *Borassus flabellifer* L. | Tad | | | | | | |
| 48. | *Boswellia serrata* Roxb. | Salayi. | | | | | | |
| 49. | *Bougainvillea spectabilis* Willd. | Bogan-wel | | | √ | | | |
| 50. | *Bridelia retusa* (L.) Spreng. | Asana | | | | | | |
| 51. | *Buchanania lanzan* Spreng. | Charoli. | | | | | | √ |
| 52. | *Butea monosperma* (Lam.) Taub. | Palas, Dhak | | | √ | | | |
| 53. | *Caesalpinia bonduc* (L.) Roxb. | Sagar-gota | | | | | | √ |
| 54. | *Caesalpinia pulcherrima* (L.) Sw. | Puraiful | | | | | | |
| 55. | *Cajanus cajan* (L.) Millsp. | Tur; Tor. | | | | √ | | √ |
| 56. | *Calotropis gigantea* (L.) R. Br. | Rui; Pandhari mandar; Arka mool | | | √ | | | |
| 57. | *Calotropis procera* (Ait.) R. Br. | Rui; Arka mool | | | √ | | | |
| 58. | *Canna indica* L. | Kardali. | | | √ | | | |
| 59. | *Cardiospermum halicacabum* L. | Kapal phodi | | | | | | |
| 60. | *Carica papaya* L. | Papai | √ | | | | | |
| 61. | *Cascabella thevetia* (L.) Lippold | Pivla Kanher | | | √ | | | |
| 62. | *Cassia absus* L. | Tarhotana-mama | √ | | | | | |
| 63. | *Cassia auriculata* L. | Aawali | | | √ | | | |
| 64. | *Cassia fistula* L. | Bahava; Baswan | | | | | | |
| 65. | *Cassia occidentalis* L. | Dukkar sheng; Ran-takla | | | | | | |
| 66. | *Cassia sophera* L. | Ran-takla | | | | | | |
| 67. | *Cassia tora* L. | Takla; Tarota; Powadya | | √ | | | | |
| 68. | *Catharanthus roseus* (L.) G.Don | Sadafuli | | | √ | | | |
| 69. | *Celastrus paniculata* Willd. | Jyotishmati; Malkangani | | | | | | |

| Sl.No. | Botanical Name | Local Name | Fruits | Leaves | Flowers | Stem | Roots | Seeds |
|---|---|---|---|---|---|---|---|---|
| 70. | *Celosia argentea* L. | Kurdu; Kirdu | | √ | | | | |
| 71. | *Centella asiatica* L. | Brahmi. | | | | | | |
| 72. | *Chlorophytum borivilianum* Sant. and Fernand. | Safed Musali | | | √ | | | |
| 73. | *Chrysanthemum indicum* L. | Shevanti | | | √ | | | |
| 74. | *Cicer arietinum* L. | Harbhara | | | | | | √ |
| 75. | *Cissampelos pareira* L. | Pahadvel; Paathaa | | | | | | |
| 76. | *Cissus quadrangularis* L | Kandvel. | | | | | | |
| 77. | *Citrullus colocynthis* Sch. | Kadu-Indrayan | | | | | | |
| 78. | *Citrullus lanatus* (Thunb.) Matsura and Nakai | Tarbuj | √ | | | | | |
| 79. | *Citrus aurantifolia* (Christm. and Panz.) Sw. | Limbu | √ | | | | | |
| 80. | *Citrus aurantium* L. | Santra | √ | | | | | |
| 81. | *Cleome gynandra* L | Pandhri Tilwan | | | | | | |
| 82. | *Cleome viscosa* L | Pivli-Tilwan | | | | | | |
| 83. | *Clitoria biflora* Dalz. | Garali | | | | | | |
| 84. | *Clitoria ternatea* L. | Gokarna | | | √ | | | |
| 85. | *Coccinia grandis* (L.) Voigt | Tondli, Tongal | √ | | | | | |
| 86. | *Cocculus hirsutus* (L.) Theob. | Para-Vel; Vasanvel | | | √ | | | |
| 87. | *Cocos nucifera* L. | Naral | √ | √ | | | | |
| 88. | *Corchorus capsularis* L. | Chuncha. | | | | | | |
| 89. | *Cordia dichotoma* Forst. *f.* | Bhokar | | | | | | |
| 90. | *Coriandrum sativum* L. | Kothimbir, Dhane | | √ | | | | √ |
| 91. | *Costus specious* (Koen.) J. E. Sm. | Pev | | | | | | |

| Sl.No. | Botanical Name | Local Name | Fruits | Leaves | Flowers | Stem | Roots | Seeds |
|---|---|---|---|---|---|---|---|---|
| 92. | *Crataeva adansonii* DC. | Waiwarna; Varuna | | √ | | | | |
| 93. | *Crinum asiaticum* L. | Gadhani-Kanda; Gadambi Kanda | | | √ | | | |
| 94. | *Cucumis melo* L. | Kharbuj. | √ | | | | | |
| 95. | *Cucumis sativus* L. | Kakdi. | √ | | | | | |
| 96. | *Cucurbita maxima* Duch. ex Lam. | Bhopla. | √ | | | | | |
| 97. | *Curculigo orchioides* Gaertn. | Kala musli; Taalmuli | | | | | | |
| 98. | *Curcuma amada* Roxb. | Ambe halad. | | | | | √ | |
| 99. | *Curcuma inodora* Blatt. | Ran halad | | | | | √ | |
| 100. | *Curcuma longa* L. | Halad | | | | | √ | |
| 101. | *Cymbopogon citratus* (DC.) Stapf | Gavti chaha | | √ | | | | |
| 102. | *Cymbopogon martinii* (Roxb.) Wats. | Rosha gavat | | √ | | | | |
| 103. | *Cynodon dactylon* (L.) Pers. | Durva. | | √ | | | | |
| 104. | *Cyperus rotundus* L | Nagarmotha | | | | | | |
| 105. | *Dalbergia sissoo* Roxb. ex DC. | Sisam | | | | | | |
| 106. | *Datura innoxia* Mill. | Pandhara dhotra. | | | √ | | | |
| 107. | *Datura metel* L. | Kala dhotra | | | √ | | | |
| 108. | *Daucus carota* L. | Gajar | | | | | √ | |
| 109. | *Delonix regia* (Boj. ex Hook.) Raf. | Gulmohar | | | | | | |
| 110. | *Desmodium gangeticum* (L.) DC. | Salvan; Shaalparni | | | | | | |
| 111. | *Diospyros melanoxylon* Roxb | Temburni | √ | | | | | |
| 112. | *Diospyros montana* Roxb | Pali | | | | | | |
| 113. | *Diplocyclos palmatus* (L.) Jeffrey | Shivalingi; Gurul | | | | | | |
| 114. | *Dolichandrone falcata* (Wall.ex DC) Seem. | Medshing; Marshing | | | √ | | | |

| Sl.No. | Botanical Name | Local Name | Fruits | Leaves | Flowers | Stem | Roots | Seeds |
|---|---|---|---|---|---|---|---|---|
| 115. | *Drimia indica* (Roxb.) Jessop | Ran Kanda | | | | | | |
| 116. | *Echinops echinatus* Roxb. | Utanti | | | | | | |
| 117. | *Eclipta prostrata* (L.) L. | Maka | | | | | | |
| 118. | *Elaeocarpus sphaericus* Gaertn. | Rudraksh | √ | | | | | |
| 119. | *Eleusine coracana* (L.) Gaertn. | Nachni | | | | | | √ |
| 120. | *Emblica officinalis* Gaertn. | Avla | √ | | | | | |
| 121. | *Enicostema axillare* (Lam.) Raynal | Chota-karait; Nai | | | | | | |
| 122. | *Euphorbia hirta* L. | Dudhi | | | | | | |
| 123. | *Evolvulus alsinoides* (L.) L. | Vishnukrant | | | | | | |
| 124. | *Ficus benghalensis* L. | Vad | √ | | | | | |
| 125. | *Ficus racemosa* L. | Umbar; Audumbar | | √ | | | | |
| 126. | *Ficus religiosa* L. | Pimpal | | | | | | |
| 127. | *Ficus virens* Ait. | Pimpri | | | | | | |
| 128. | *Flacourtia indica* (Burm. *f.*) Merr. | Atrun | | | | | | |
| 129. | *Garnga pinnata* Roxb. | Kakad, Kakod | | | | | | |
| 130. | *Gloriosa superba* L. | Kal-lawi | | | √ | | | |
| 131. | *Gmelina arborea* Roxb. | Shivan | | | | | | |
| 132. | *Gossypium arboreum* L. | Devkapas | √ | | | | | |
| 133. | *Gossypium herbaceum* L. | Kapas; Kapus | √ | | | | | |
| 134. | *Haldina cordifolia* (Roxb.) Ridsd. | Haldu | | | | | | |
| 135. | *Hardwickia binata* Roxb. | Anjan | | | | | | |
| 136. | *Helicteres isora* L. | Murudsheng | | | | | | |
| 137. | *Hemidesmus indicus* (L.) Schult. | Sariva; Anantamul | | | | | √ | |
| 138. | *Hibiscus rosa-sinensis* L. | Jasvand | | | √ | | | |
| 139. | *Hibiscus sabdariffa* L. | Lal ambadi | | | √ | | | |

| Sl.No. | Botanical Name | Local Name | Fruits | Leaves | Flowers | Stem | Roots | Seeds |
|---|---|---|---|---|---|---|---|---|
| 140. | *Holarrhena pubescens* (Buch.-Ham.) Wall. ex G. Don | Dahikudi | | | √ | | | |
| 141. | *Holoptelea integrifolia* Planch. | Vavla | | | | | | |
| 142. | *Hymenodictyon ovixense* (Roxb.) Mabberly | Madala | | | | | | |
| 143. | *Ipomoea illustris* (Clarke) Prain. | Kumrhoney | | | | | | |
| 144. | *Ipomoea nil* (L.) Roth | Nil-pushpa | | | | | | |
| 145. | *Ixora coccinea* L. | Bokadi | | | √ | | | |
| 146. | *Jasminum officinale* L. | Chameli. | | | √ | | | |
| 147. | *Jasminum sambac* (L.) Ait. | Mogra | | | √ | | | |
| 148. | *Jatropha curcas* L. | Moghli errand | | | | | | |
| 149. | *Justicia adhatoda* L. | Adulasa | | √ | | | | |
| 150. | *Kalanchoe pinnata* (Lam.) Pers. | Panphuti | | | | | | |
| 151. | *Kohautia aspera* (Heyne ex Roth) Brem. | Khetpara | | | | | | |
| 152. | *Lagenaria siceraria* (Mol.) Standl. | Dudhi | √ | | | | | |
| 153. | *Lagerstroemia microcarpa* Wight | Nana Bondara; Hina | | | | | | |
| 154. | *Lantana camara* L. | Ghaneri, Gansubai. | | | | | | |
| 155. | *Lawsonia inermis* L. | Kadu mendi; Mehandi | | √ | | | | |
| 156. | *Leucas cephalotes* (Roth) Spr. | Ghumma. | | | | | | |
| 157. | *Limonia acidissima* L. | Kavath | √ | | | | | |
| 158. | *Luffa acutangula* (L.) Roxb. | Shirale | √ | | | | | |
| 159. | *Madhuca longifolia* (Koen.) Mac Bride | Moha, Mahu, Mahuda | √ | | √ | | | |
| 160. | *Mallotus philippensis* (Lam.) Muell-Arg. | Shendri | √ | | | | | |

| Sl.No. | Botanical Name | Local Name | Fruits | Leaves | Flowers | Stem | Roots | Seeds |
|---|---|---|---|---|---|---|---|---|
| 161. | *Mangifera indica* L. | Amba. | √ | | √ | | | |
| 162. | *Martynia annua* L. | Winchu. | √ | | | | | |
| 163. | *Melia azedarach* L. | Bakana neem | | | | | | |
| 164. | *Melilotus alba* Medik. ex Desr. | Ran methi | | √ | | | | |
| 165. | *Merremia dissecta* (Jacq.) Hallier *f.* | | | | | | | |
| 166. | *Michelia champaca* L. | Sonchapha | | | √ | | | |
| 167. | *Mimosa hamata* Willed. | Arati | | | | | | |
| 168. | *Mimusops elengi* L. | Bakul. | | | √ | | | |
| 169. | *Mirabilis jalapa* L. | Gulbakshi | | | √ | | | |
| 170. | *Mitragyna parvifolia* (Roxb.) Korth. | | | | | √ | | |
| 171. | *Momordica dioica* Roxb. ex Willd. | Kartoli; Katurle | √ | | | | | |
| 172. | *Moringa oleifera* Lam. | Shevga | √ | | | | | |
| 173. | *Mucuna pruriens* (L.) DC. | Kahj-kuili | | | | | | |
| 174. | *Mukia maderaspatana* L. | Ankhphod | | | | | | |
| 175. | *Musa paradisiaca* L. | Kel | √ | √ | √ | | | |
| 176. | *Nelumbo nucifera* Gaertn. | Kamal | | | √ | | | |
| 177. | *Nerium indicum* Mill. | Kanher | | | √ | | | |
| 178. | *Nyctanthes arbortristis* L. | Parijatak | | | √ | | | |
| 179. | *Nymphaea nouchali* Burm.*f.* | Kamal; Kamay | | | √ | | | |
| 180. | *Nymphaea pubescens* Willd | Kamal | | | √ | | | |
| 181. | *Ocimum basilicum* L. | Sabja | | √ | √ | | | |
| 182. | *Ocimum gratissimum* L. | Ram Tulasi | | √ | √ | | | |
| 183. | *Ocimum tenuiflorum L.* | Krishnatulas | | √ | | | | |
| 184. | *Opuntia elatior* Mill. | Nivdung | | | | | | |
| 185. | *Oroxylum indicum* (L.) Vent. | Tetu. | | | | | | |

| Sl.No. | Botanical Name | Local Name | Fruits | Leaves | Flowers | Stem | Roots | Seeds |
|---|---|---|---|---|---|---|---|---|
| 186. | *Oryza sativa* L. | Tandul, Dhan | | | | | | √ |
| 187. | *Oxalis corniculata* L. | Ambuti | | | | | | |
| 188. | *Pandanus odoratissimus* L. *f.* | Kevada | | √ | | | | |
| 189. | *Passiflora edulis* Sims | Krishna Kamal | | | √ | | | |
| 190. | *Pedalium murex* L. | Mothi-Gokhru, Ubha Gokharu | | | | | | |
| 191. | *Pennisetum americanum* L. | Bajaras, Bajari | | | | | | √ |
| 192. | *Pergularia daemia* (Forssk.) Chiov. | Utaran; Rota | | | | | | |
| 193. | *Phoenix dactylifera* (L) Roxb. | Khajur | √ | | | | | |
| 194. | *Phyla nodiflora* (L.) Greene | Mundi | | | | | | |
| 195. | *Phyllanthus acidus* (L.) K. Skeels | Rai Awla | √ | | | | | |
| 196. | *Phyllanthus amarus* L. | Bhui Awala | √ | | | | | |
| 197. | *Physalis micrantha* L. | Koyi | | | | | | |
| 198. | *Piper betle* L. | Paan | | √ | | | | |
| 199. | *Plumbago zeylanica* L. | Chitrak | | | | | | |
| 200. | *Plumeria rubra* L. forma *acuminata* (Ait.) Santapaul and Irani ex Shah | Pandhara Chapha | | √ | √ | | | |
| 201. | *Polyalthia longifolia* Sonn. | Ashok | | √ | | | | |
| 202. | *Pongamia pinnata* (L.) Pierre | Karanj | | | √ | | | |
| 203. | *Prosopis cineraria* (L.) Druce | Saundad, Samadi | | | | | | |
| 204. | *Psidium guajava* L. | Peru | √ | | | | | |
| 205. | *Pterocarpus marsupium* Roxb. | Bibla; Bijjaka | √ | | | | | |
| 206. | *Pueraria tuberosa* (Roxb. ex Willd.) DC. | Bhui Kohala | √ | | | | | |
| 207. | *Punica granatum* L. | Dalimb | √ | | | | | |
| 208. | *Quisqualis indica* L. | Madhumalti | | | √ | | | |

| Sl.No. | Botanical Name | Local Name | Fruits | Leaves | Flowers | Stem | Roots | Seeds |
|---|---|---|---|---|---|---|---|---|
| 209. | *Raphanus sativus* L. | Mula | | √ | | | √ | |
| 210. | *Ricinus communis* L. | Erand | | | | √ | | |
| 211. | *Rosa damascena* Mill. | Gulab | | | √ | | | |
| 212. | *Saccharum officinarum* L. | Uss | | | | √ | | |
| 213. | *Salvadora persica* L. | Pilu | √ | | | | | |
| 214. | *Santalum album* L. | Chandan | | | | √ | | |
| 215. | *Sapindus emarginatus* Vahl. | Ritha | | | | | | |
| 216. | *Saraca asoca* (Roxb.) de Wilde | Samudra Ashok | | √ | | | | |
| 217. | *Schleichera oleosa* (Lour.) Oken | Kusum | | √ | | | | |
| 218. | *Securinega virosa* (Roxb. ex Willd.) Baill. | Pandharphali | √ | | | | | |
| 219. | *Semecarpus anacardium* L. | Bhilava | √ | | | | | |
| 220. | *Sesamum orientale* L. | Til | | | | | | √ |
| 221. | *Sesbania grandiflora* (L.) Poir. | Hadga | | | | | | |
| 222. | *Sesbania sesban* (L.) Merr. | Sheuri, Shewari | | | | | | |
| 223. | *Setaria italica* (L.) P. Beauv. | Bhandi | | | | | | |
| 224. | *Sida acuta* Burm. *f.* | Bokya | | | | | | |
| 225. | *Sida cordata* (Burm.*f.*) Borss. | Bala | | | | | | |
| 226. | *Sida spinosa* L. | Gangeti; Kantalobata | | | | | | |
| 227. | *Solanum anguivi* Lam. | Mothi ringni | √ | | | | | |
| 228. | *Solanum melongena* L. | Vangi | √ | | | | | |
| 229. | *Solanum nigrum* L. | Kangani | | | | | | |
| 230. | *Solanum virginianum* L. | Bhuiringani. | | | | | | |
| 231. | *Sorghum vulgare* Pers. | Jowar. | | | | | | √ |
| 232. | *Spilanthus calva* DC. | Akkalkadha | | | √ | | | |

| Sl.No. | Botanical Name | Local Name | Fruits | Leaves | Flowers | Stem | Roots | Seeds |
|---|---|---|---|---|---|---|---|---|
| 233. | *Spondias pinnata* (L. *f.*) Kurz. | Ambada | | | | | | |
| 234. | *Sterculia urens* Roxb. | Kadai; Kadoni; Kevda; Kudal | | | | | | |
| 235. | *Syzygium cumini* (L.) Skeels | Jambhul. | √ | | | | | |
| 236. | *Tabernaemontana divaricata* (L.) R. Br. | Tagar; Chandani | √ | | | | | |
| 237. | *Tagetes erecta* L. | Zendu | √ | | | | | |
| 238. | *Tamarindus indica* L. | Imli | √ | | | | | |
| 239. | *Tectona grandis* L. | Sag | | √ | | √ | | |
| 240. | *Tephrosia purpurea* Pers. | Sarpankha | | | | | | |
| 241. | *Terminalia arjuna* (Roxb. ex DC.) Wight and Arn. | Arjun | | | | | | |
| 242. | *Terminalia bellirica* (Gaertn.) Roxb. | Beheda | √ | | | | | |
| 243. | *Thespesia populnea* (L.) Soland. ex Corr. | Paras pimpal | | | | | | |
| 244. | *Thuja plicata* ex D. Don | Vidya | | √ | | | | |
| 245. | *Tinospora cordifolia* (Willd.) Miers ex Hook. *f.* and Thoms | Gudvel | | | | | | |
| 246. | *Tribulus terrestris* L. | Gokhru; Gokshura mool | | | | | | |
| 247. | *Tricholepis indica* (Willd.) Almeida | Dahan | | | | | | |
| 248. | *Trichosanthes cucumerina* L. | Jangli padval | | | | | | |
| 249. | *Trichosanthes tricuspidata* Lour. | Mukul | | | | | | |
| 250. | *Tricticum aestivum* L. | Gahu | | | | | | √ |
| 251. | *Tridax procumbens* L. | Dagdi pala | | | | | | |
| 252. | *Vigna mungo* (L.) Hepper | Udid | | | | | | √ |
| 253. | *Vigna radiata* (L.) R. Wilczek | Sona mug | | | | | | √ |
| 254. | *Vigna trilobata* (L.) Verdc. | Mung | | | | | | √ |

| Sl.No. | Botanical Name | Local Name | Fruits | Leaves | Flowers | Stem | Roots | Seeds |
|---|---|---|---|---|---|---|---|---|
| 255. | *Vitex negundo* L. | Nirgundi | | √ | | | | |
| 256. | *Vitis vinifera* L. | Draksha | √ | | | | | |
| 257. | *Withania somnifera* (L.) Dunal | Ashwagandha | | | | | | |
| 258. | *Woodfordia fruticosa* (L.) Kurz. | Dhoneyti | | | | | | |
| 259. | *Wrightia tinctoria* R.Br. | Pandhra kuda | | | | | | |
| 260. | *Xanthium indicum* Koen. | Sankeshwar | | | | | | |
| 261. | *Zea mays* L. | Maka, Makki, Makai | | | | | | √ |
| 262. | *Zingiber officinale* Rosc. | Adrak | | | | | √ | |
| 263. | *Zizyphus mauritiana* Lam. | Bor | √ | | | | | |

**Table 5: Plant Parts Used for Medicinal Purposes**

| Sl.No. | Botanical Name | Local Name | Fruits | Leaves | Flowers | Stem | Roots | Seeds |
|---|---|---|---|---|---|---|---|---|
| 1. | *Abelmoschus esculentus* (L.) Moench. | Bhendi | √ | | | √ | √ | √ |
| 2. | *Abelmoschus manihot* (L.) Medik. | Kasturi bhendi, Ran bhendi | √ | | | | | |
| 3. | *Abitulon indicum* (L.) | Sikka, Atti. | √ | √ | | √ | √ | |
| 4. | *Abrus precatorius* L. | Gunj, Ratti | √ | | | | √ | √ |
| 5. | *Acacia chundra* (Roxb. ex Rottle.) Willd. | Khair | | | | | √ | |
| 6. | *Acacia ferruginea* DC. | Devbhabul | √ | | | √ | √ | |
| 7. | *Acacia nilotica* (L.) Del. | Ramkathi. | √ | | | √ | | √ |
| 8. | *Acalypha indica* L. | Kokali | | √ | | | | |
| 9. | *Achyranthes aspera* L. | Aghada, Chirchir | | √ | | √ | | |
| 10. | *Acorus calamus* L. | Vekhand; Yekhand | | | | √ | | |
| 11. | *Adansonia digitata* L. | Gorakh chinch | √ | | | | | |
| 12. | *Aegle marmelos* (L.) Corr. | Bel phal, Bel mool, Bel. | √ | √ | | | | |
| 13. | *Ailanthus excelsa* Roxb. | Maharuk | | √ | | √ | | |
| 14. | *Albizia lebbeck* (L.) Willd. | Shirish | | √ | | √ | | |
| 15. | *Allium cepa* L. | Kanda | | | | | √ | |
| 16. | *Allium sativum* L. | Lasun | | | | | √ | |
| 17. | *Aloe vera* L. | Korphad | | | | | | |
| 18. | *Alstonia scholaris* R.Br. | Saptaparn | | | | | √ | |
| 19. | *Amaranthus hybridus* L | Rajgira | | | | | | √ |
| 20. | *Amaranthus spinosus* L. | Kate – math | | √ | | | √ | |
| 21. | *Ammannia baccifera* L. | Aagia | | √ | | | | |
| 22. | *Andrographis echioides* (L.) Nees. | Kalmegh | | √ | | | | |

| Sl.No. | Botanical Name | Local Name | Fruits | Leaves | Flowers | Stem | Roots | Seeds |
|---|---|---|---|---|---|---|---|---|
| 23. | *Anethum graveolens* L. | Balant shepu | √ | √ | √ | √ | √ | √ |
| 24. | *Anisomeles malabarica* R. Br. | Kalatumba | | | | | | |
| 25. | *Annona reticulata* L. | Ramphal. | | | | | | |
| 26. | *Annona squamosa* L. | Sitaphal | √ | √ | | | √ | √ |
| 27. | *Anogeissus latifolia* Wall. | Dhavda | | | | | √ | |
| 28. | *Apium graveolens* L. | Shepu | | | | | | √ |
| 29. | *Arachis hypogaea* L. | Bhuimug | | | | | | √ |
| 30. | *Argemone mexicana* L. | Pivla Dhotra | √ | | | √ | | √ |
| 31. | *Argyreia nervosa* (Burm. *f.*) Bojer | Samudrashok | | √ | | √ | | |
| 32. | *Aristolochia bracteolata* Lam. | Kidamar | | | | √ | | |
| 33. | *Asparagus racemosus* Willd | Shatavari, Chinyar | | √ | | | √ | |
| 34. | *Azadirachta indica* A. Juss. | Kadu neem | | | | | √ | |
| 35. | *Bacopa monnieri* Penn. | Brahmi | √ | √ | √ | √ | √ | √ |
| 36. | *Balanitis aegyptiaca* (L.) Delile | Hinganbet | √ | | | | √ | √ |
| 37. | *Baliospermum raziana* Keshav *et.* Yog | Danti | | √ | | | √ | √ |
| 38. | *Bambusa arundinacea* (Retz.)Willd. | Kalak; Bambu. | | | | | | √ |
| 39. | *Barleria lawii* L. | Akhar | | | | | | |
| 40. | *Barleria prionitis* L. | Kate Koranti. | | √ | √ | | √ | |
| 41. | *Bauhinia racemosa* Lamk. | Apta | | √ | | | | |
| 42. | *Bauhinia tomentosa* L. | Ashundar | | | | | √ | |
| 43. | *Bauhinia variegata* L. | Kanchan. | √ | | | | √ | |
| 44. | *Benincasa hispida* (Thunb.) Cogn | Kohla. | √ | | | | | |
| 45. | *Boerhavia repens* L. var. *diffusa* (L.) Hook *f.* | Punarnawa | | | | | √ | |

| Sl.No. | Botanical Name | Local Name | Fruits | Leaves | Flowers | Stem | Roots | Seeds |
|---|---|---|---|---|---|---|---|---|
| 46. | *Bombax ceiba* L. | Kate Sawar, Hawar | | √ | | | √ | |
| 47. | *Borassus flabellifer* L. | Tad | √ | | | | | |
| 48. | *Boswellia serrata* Roxb. | Salayi. | √ | | | | √ | √ |
| 49. | *Bougainvillea spectabilis* Willd. | Bogan-wel | | | | | | |
| 50. | *Bridelia retusa* (L.) Spreng. | Asana | | | | | √ | |
| 51. | *Buchanania lanzan* Spreng. | Charoli. | | | | | | √ |
| 52. | *Butea monosperma* (Lam.) Taub. | Palas, Dhak | √ | | | | √ | |
| 53. | *Caesalpinia bonduc* (L.) Roxb. | Sagar-gota | | √ | | | √ | √ |
| 54. | *Caesalpinia pulcherrima* (L.) Sw. | Puraiful | | | | | | √ |
| 55. | *Cajanus cajan* (L.) Millsp. | Tur; Tor. | | √ | | | | √ |
| 56. | *Calotropis gigantea* (L.) R. Br | Rui; Pandhari mandar; Arka mool | | | | | √ | |
| 57. | *Calotropis procera* (Ait.) R. Br. | Rui; | | | | | | |
| 58. | *Canna indica* L. | Kardali. | | | √ | | √ | |
| 59. | *Cardiospermum halicacabum* L. | Kapal phodi | √ | √ | √ | √ | √ | |
| 60. | *Carica papaya* L. | Papai | √ | | | | | |
| 61. | *Cascabella thevetia* (L.) Lippold | Pivla Kanher | | | | | | √ |
| 62. | *Cassia absus* L. | Tarhotana-mama | | √ | √ | | | |
| 63. | *Cassia auriculata* L | Aawali | | | | | √ | |
| 64. | *Cassia fistula* L. | Bahava; Baswan | | √ | √ | √ | √ | |
| 65. | *Cassia occidentalis* L. | Dukkar sheng; Ran-takla | | | √ | | √ | |
| 66. | *Cassia sophera* L. | Ran-takla | | √ | | | | |
| 67. | *Cassia tora* L. | Takla; Tarota; Powadya | | √ | √ | | | √ |
| 68. | *Catharanthus roseus* (L.) G.Don | Sadafuli | √ | √ | √ | √ | √ | |
| 69. | *Celastrus paniculata* Willd. | Jyotishmati; Malkangani | | | | | | √ |

| Sl.No. | Botanical Name | Local Name | Fruits | Leaves | Flowers | Stem | Roots | Seeds |
|---|---|---|---|---|---|---|---|---|
| 70. | *Celosia argentea* L. | Kurdu; Kirdu | √ | √ | √ | √ | √ | |
| 71. | *Centella asiatica* L. | Brahmi. | √ | √ | √ | √ | √ | |
| 72. | *Chlorophytum borivilianum* Sant. and Fernand. | Safed Musali | | | | | √ | |
| 73. | *Chrysanthemum indicum* L. | Shevanti | | | | | | |
| 74. | *Cicer arietinum* L. | Harbhara | | | | | | √ |
| 75. | *Cissampelos pareira* L. | Pahadvel; Paathaa | | | | | √ | |
| 76. | *Cissus quadrangularis* L | Kandvel. | | √ | | √ | √ | |
| 77. | *Citrullus colocynthis* Sch. | Kadu-Indrayan | √ | | | | √ | |
| 78. | *Citrullus lanatus* (Thunb.) Matsura and Nakai | Tarbuj | | | | | | |
| 79. | *Citrus aurantifolia* (Christm. and Panz.) Sw. | Limbu | √ | √ | | | | |
| 80. | *Citrus aurantium* L. | Santra | √ | | | | | |
| 81. | *Cleome gynandra* L | Pandhri Tilwan | √ | √ | √ | √ | √ | |
| 82. | *Cleome viscosa* L | Pivli-Tilwan | √ | √ | | | | |
| 83. | *Clitoria biflora* Dalz. | Garali | | | | | | |
| 84. | *Clitoria ternatea* L. | Gokarna | | | √ | | | |
| 85. | *Coccinia grandis* (L.) Voigt | Tondli, Tongal | √ | | | | √ | |
| 86. | *Cocculus hirsutus* (L.) Theob. | Para-Vel; Vasanvel | | √ | | | | |
| 87. | *Cocos nucifera* L. | Naral | √ | | | | | |
| 88. | *Corchorus capsularis* L. | Chuncha. | | √ | | | | |
| 89. | *Cordia dichotoma* Forst. *f.* | Bhokar | √ | | | | √ | |
| 90. | *Coriandrum sativum* L. | Kothimbir, Dhane | | | | | | √ |
| 91. | *Costus specious* (Koen.) J. E. Sm. | Pev | | | | √ | √ | |

| Sl.No. | Botanical Name | Local Name | Fruits | Leaves | Flowers | Stem | Roots | Seeds |
|---|---|---|---|---|---|---|---|---|
| 92. | *Crataeva adansonii* DC. | Waiwarna; Varuna | | | | √ | | |
| 93. | *Crinum asiaticum* L. | Gadhani-Kanda; Gadambi Kanda | | √ | | | √ | |
| 94. | *Cucumis melo* L. | Kharbuj. | √ | | | | | |
| 95. | *Cucumis sativus* L. | Kakdi. | √ | √ | | | | |
| 96. | *Cucurbita maxima* Duch. ex Lam. | Bhopla. | √ | | | | | |
| 97. | *Curculigo orchioides* Gaertn. | Kala musli; Taalmuli | | | | | √ | |
| 98. | *Curcuma amada* Roxb. | Ambe halad. | | | | | √ | |
| 99. | *Curcuma inodora* Blatt. | Ran halad | | | | | √ | |
| 100. | *Curcuma longa* L. | Halad | | | | | √ | |
| 101. | *Cymbopogon citratus* (DC.) Stapf | Gavti chaha | | | | | | |
| 102. | *Cymbopogon martinii* (Roxb.) Wats. | Rosha gavat | | √ | | | | |
| 103. | *Cynodon dactylon* (L.) Pers. | Durva. | | √ | | | √ | |
| 104. | *Cyperus rotundus* L | Nagarmotha | | | | | √ | |
| 105. | *Dalbergia sissoo* Roxb. ex DC. | Sisam | | √ | | | | |
| 106. | *Datura innoxia* Mill. | Pandhara dhotra. | √ | √ | | | | |
| 107. | *Datura metel* L. | Kala dhotra | | √ | | | √ | √ |
| 108. | *Daucus carota* L. | Gajar | | √ | | | √ | |
| 109. | *Delonix regia* (Boj. ex Hook.) Raf. | Gulmohar | | | | | √ | |
| 110. | *Desmodium gangeticum* (L.) DC. | Salvan; Shaalparni | | | | | √ | |
| 111. | *Diospyros melanoxylon* Roxb | Temburni | √ | | | √ | | |
| 112. | *Diospyros montana* Roxb | Pali | | | √ | | | |
| 113. | *Diplocyclos palmatus* (L.) Jeffrey | Shivalingi; Gurul | | √ | | | | |
| 114. | *Dolichandrone falcata* (Wall.ex DC) Seem. | Medshing; Marshing | | | | | | |

| Sl.No. | Botanical Name | Local Name | Fruits | Leaves | Flowers | Stem | Roots | Seeds |
|---|---|---|---|---|---|---|---|---|
| 115. | *Drimia indica* (Roxb.) Jessop | Ran Kanda | | | | | √ | |
| 116. | *Echinops echinatus* Roxb. | Utanti | √ | √ | √ | √ | √ | √ |
| 117. | *Eclipta prostrata* (L.) L. | Maka | √ | √ | √ | √ | √ | √ |
| 118. | *Elaeocarpus sphaericus* Gaertn. | Rudraksh | √ | | | | | |
| 119. | *Eleusine coracana* (L.) Gaertn. | Nachni | | | | | | √ |
| 120. | *Emblica officinalis* Gaertn. | Avla | √ | | | | | |
| 121. | *Enicostema axillare* (Lam.) Raynal | Chota-karait; Nai | √ | √ | √ | √ | √ | |
| 122. | *Euphorbia hirta* L. | Dudhi | √ | √ | √ | √ | √ | |
| 123. | *Evolvulus alsinoides* (L.) L. | Vishnukrant | √ | √ | √ | √ | √ | |
| 124. | *Ficus benghalensis* L. | Vad | | √ | | | √ | |
| 125. | *Ficus racemosa* L. | Umbar; Audumbar | √ | | | | √ | |
| 126. | *Ficus religiosa* L. | Pimpal | √ | √ | | | | |
| 127. | *Ficus virens* Ait. | Pimpri | | √ | | | | |
| 128. | *Flacourtia indica* (Burm. *f.*) Merr. | Atrun | √ | | | √ | √ | |
| 129. | *Garnga pinnata* Roxb. | Kakad, Kakod | | | | | | |
| 130. | *Gloriosa superba* L. | Kal-lawi | | √ | | | √ | |
| 131. | *Gmelina arborea* Roxb. | Shivan | | | | √ | √ | |
| 132. | *Gossypium arboreum* L. | Devkapas | | | | | | |
| 133. | *Gossypium herbaceum* L. | Kapas; Kapus | | | | | | √ |
| 134. | *Haldina cordifolia* (Roxb.) Ridsd. | Haldu | | | | | | |
| 135. | *Hardwickia binata* Roxb. | Anjan | | | | √ | | |
| 136. | *Helicteres isora* L. | Murudsheng | √ | | | √ | √ | √ |
| 137. | *Hemidesmus indicus* (L.) Schult. | Sariva; Anantamul | | | | | √ | |
| 138. | *Hibiscus rosa-sinensis* L. | Jasvand | | √ | | | √ | |
| 139. | *Hibiscus sabdariffa* L. | Lal ambadi | | √ | | | | |

| Sl.No. | Botanical Name | Local Name | Fruits | Leaves | Flowers | Stem | Roots | Seeds |
|---|---|---|---|---|---|---|---|---|
| 140. | *Holarrhena pubescens* (Buch.-Ham.) Wall. ex G. Don | Dahikudi | | | | √ | | √ |
| 141. | *Holoptelea integrifolia* Planch. | Papda | | √ | | √ | | |
| 142. | *Hymenodictyon ovixense* (Roxb.) Mabberly | Madala | | | | √ | √ | |
| 143. | *Ipomoea illustris* (Clarke) Prain. | Kumrhoney | | | | | √ | |
| 144. | *Ipomoea nil* (L.) Roth | Nil-pushpa | √ | √ | √ | √ | √ | |
| 145. | *Ixora coccinea* L. | Bokadi | | | | | | |
| 146. | *Jasminum officinale* L. | Chameli. | √ | | | | | |
| 147. | *Jasminum sambac* (L.) Ait. | Mogra | | √ | √ | | | |
| 148. | *Jatropha curcas* L. | Moghli errand | | | | √ | | √ |
| 149. | *Justicia adhatoda* L. | Adulasa | | √ | | | | |
| 150. | *Kalanchoe pinnata* (Lam.) Pers. | Panphuti | | √ | | | | |
| 151. | *Kohautia aspera* (Heyne ex Roth) Brem. | Khetpara | √ | √ | √ | √ | √ | |
| 152. | *Lagenaria siceraria* (Mol.) Standl. | Dudhi | √ | | | | | |
| 153. | *Lagerstroemia microcarpa* Wight | Nana Bondara; Hina | | | | √ | | |
| 154. | *Lantana camara* L. | Ghaneri, Gansubai. | | √ | | | | |
| 155. | *Lawsonia inermis* L. | Kadu mendi; Mehandi | | √ | | | | √ |
| 156. | *Leucas cephalotes* (Roth) Spr. | Ghumma. | | √ | | | | |
| 157. | *Limonia acidissima* L. | Kavath | √ | | | √ | √ | |
| 158. | *Luffa acutangula* (L.) Roxb. | Shirale | √ | | | | √ | |
| 159. | *Madhuca longifolia* (Koen.) Mac Bride | Moha, Mahu, Mahuda | | √ | √ | √ | | √ |
| 160. | *Mallotus philippensis* (Lam.) Muell-Arg. | Shendri | √ | | | | | √ |

| Sl.No. | Botanical Name | Local Name | Fruits | Leaves | Flowers | Stem | Roots | Seeds |
|---|---|---|---|---|---|---|---|---|
| 161. | *Mangifera indica* L. | Amba. | √ | √ | | √ | | |
| 162. | *Martynia annua* L. | Winchu. | √ | √ | | | | √ |
| 163. | *Melia azedarach* L. | Bakana neem | | √ | | | | |
| 164. | *Melilotus alba* Medik. ex Desr. | Ran methi | | | | | | |
| 165. | *Merremia dissecta* (Jacq.) Hallier *f.* | | | | | | | |
| 166. | *Michelia champaca* L. | Sonchapha | | √ | | | | |
| 167. | *Mimosa hamata* Willed. | Arati | | | | | | |
| 168. | *Mimusops elengi* L. | Bakul. | | | | √ | | |
| 169. | *Mirabilis jalapa* L. | Gulbakshi | | √ | | | | |
| 170. | *Mitragyna parvifolia* (Roxb.) Korth. | | | | | | | |
| 171. | *Momordica dioica* Roxb. ex Willd. | Kartoli; Katurle | √ | | | | √ | |
| 172. | *Moringa oleifera* Lam. | Shevga | | | √ | √ | √ | √ |
| 173. | *Mucuna pruriens* (L.) DC. | Kahj-kuili | | √ | | | √ | √ |
| 174. | *Mukia maderaspatana* L. | Ankhphod | | √ | | | | |
| 175. | *Musa paradisiaca* L. | Kel | √ | √ | | | √ | |
| 176. | *Nelumbo nucifera* Gaertn. | Kamal | | √ | | | | √ |
| 177. | *Nerium indicum* Mill. | Kanher | | √ | √ | | | |
| 178. | *Nyctanthes arbortristis* L. | Parijatak | | √ | √ | | | |
| 179. | *Nymphaea nouchali* Burm.*f.* | Kamal; Kamay | | | √ | | | |
| 180. | *Nymphaea pubescens* Willd | Kamal | | | √ | | | |
| 181. | *Ocimum basilicum* L. | Sabja | | √ | | | | √ |
| 182. | *Ocimum gratissimum* L. | Ram Tulasi | | √ | | | | |
| 183. | *Ocimum tenuiflorum L.* | Krishnatulas | | √ | | | | |
| 184. | *Opuntia elatior* Mill. | Nivdung | | | | √ | | |
| 185. | *Oroxylum indicum* (L.) Vent. | Tetu. | | √ | | √ | | |

| Sl.No. | Botanical Name | Local Name | Fruits | Leaves | Flowers | Stem | Roots | Seeds |
|---|---|---|---|---|---|---|---|---|
| 186. | *Oryza sativa* L. | Tandul, Dhan | | | | | | √ |
| 187. | *Oxalis corniculata* L. | Ambuti | | √ | | | | |
| 188. | *Pandanus odoratissimus* L. *f.* | Kevada | | | √ | | | |
| 189. | *Passiflora edulis* Sims | Krishna Kamal | | | | | | |
| 190. | *Pedalium murex* L. | Mothi-Gokhru, Ubha Gokharu | √ | | | | | |
| 191. | *Pennisetum americanum* L. | Bajaras, Bajari | | | | | | √ |
| 192. | *Pergularia daemia* (Forssk.) Chiov. | Utaran; Rota | √ | √ | √ | √ | √ | |
| 193. | *Phoenix dactylifera* (L) Roxb. | Khajur | √ | | | | | |
| 194. | *Phyla nodiflora* (L.) Greene | Mundi | √ | √ | √ | √ | √ | |
| 195. | *Phyllanthus acidus* (L.) K. Skeels | Rai Awala | | | | | | |
| 196. | *Phyllanthus amarus* L. | Bhui Awla | √ | √ | √ | √ | √ | |
| 197. | *Physalis micrantha* L. | Koyi | | √ | | | | |
| 198. | *Piper betle* L. | Paan | | √ | | | | |
| 199. | *Plumbago zeylanica* L. | Chitrak | | √ | | √ | √ | |
| 200. | *Plumeria rubra* L. forma *acuminata* (Ait.) Santapaul and Irani ex Shah | Pandhara Chapha | | √ | | √ | | |
| 201. | *Polyalthia longifolia* Sonn. | Ashok | | | | | | |
| 202. | *Pongamia pinnata* (L.) Pierre | Karanj | √ | √ | | | | √ |
| 203. | *Prosopis cineraria* (L.) Druce | Saundad, Samadi | | | | | | |
| 204. | *Psidium guajava* L. | Peru | √ | √ | | | | |
| 205. | *Pterocarpus marsupium* Roxb. | Bibla; Bijjaka | | √ | | √ | | |
| 206. | *Pueraria tuberosa* (Roxb. ex Willd.) DC. | Bhui Kohala | | | | | √ | |
| 207. | *Punica granatum* L. | Dalimb | √ | √ | | | | √ |
| 208. | *Quisqualis indica* L. | Madhumalti | | √ | √ | | | |

| Sl.No. | Botanical Name | Local Name | Fruits | Leaves | Flowers | Stem | Roots | Seeds |
|---|---|---|---|---|---|---|---|---|
| 209. | *Raphanus sativus* L. | Mula | | | | | √ | |
| 210. | *Ricinus communis* L. | Erand | | √ | | | | √ |
| 211. | *Rosa damascena* Mill | Gulab | | | | | | |
| 212. | *Saccharum officinarum* L. | Uss | | | | √ | | |
| 213. | *Salvadora persica* L. | Pilu | | √ | | √ | | √ |
| 214. | *Santalum album* L. | Chandan | | | | √ | | |
| 215. | *Sapindus emarginatus* Vahl. | Ritha | | | | √ | | |
| 216. | *Saraca asoca* (Roxb.) de Wilde | Samudra Ashok | | √ | √ | | | |
| 217. | *Schleichera oleosa* (Lour.) Oken | Kusum | √ | | | √ | | √ |
| 218. | *Securinega virosa* (Roxb. ex Willd.) Baill. | Pandharphali | √ | √ | | | | |
| 219. | *Semecarpus anacardium* L. | Bhilava | √ | | | | | √ |
| 220. | *Sesamum orientale* L. | Til | | | | | | √ |
| 221. | *Sesbania grandiflora* (L.) Poir. | Hadga | | | | | | |
| 222. | *Sesbania sesban* (L.) Merr. | Sheuri, Shewari | | √ | √ | | | |
| 223. | *Setaria italica* (L.) P. Beauv. | Bhandi | | | | | | |
| 224. | *Sida acuta* Burm. *f.* | Bokya | √ | √ | | | √ | |
| 225. | *Sida cordata* (Burm.*f.*) Borss. | Bala | √ | √ | √ | √ | √ | |
| 226. | *Sida spinosa* L. | Gangeti; Kantalobata | | √ | | | √ | |
| 227. | *Solanum anguivi* Lam. | Mothi ringni | | | | | √ | |
| 228. | *Solanum melongena* L. | Vangi | | | | | | |
| 229. | *Solanum nigrum* L. | Kangani | √ | √ | √ | √ | √ | |
| 230. | *Solanum virginianum* L. | Bhuiringani. | √ | √ | √ | √ | √ | |
| 231. | *Sorghum vulgare* Pers. | Jowar. | | | | | | |
| 232. | *Spilanthus calva* DC. | Akkalkadha | | | √ | | √ | |

| Sl.No. | Botanical Name | Local Name | Fruits | Leaves | Flowers | Stem | Roots | Seeds |
|---|---|---|---|---|---|---|---|---|
| 233. | *Spondias pinnata* (L. *f.*) Kurz. | Ambada | | √ | | √ | | |
| 234. | *Sterculia urens* Roxb. | Kadai; Kadoni; Kevda; Kudal | Gum | | | | | |
| 235. | *Syzygium cumini* (L.) Skeels | Jambhul. | √ | √ | | | | √ |
| 236. | *Tabernaemontana divaricata* (L.) R. Br. | Tagar; Chandani | | √ | | | | |
| 237. | *Tagetes erecta* L. | Zendu | | √ | | | | |
| 238. | *Tamarindus indica* L. | Imli | √ | √ | | | | √ |
| 239. | *Tectona grandis* L. | Sag | | | | √ | | √ |
| 240. | *Tephrosia purpurea* Pers. | Sarpankha | | √ | √ | √ | √ | |
| 241. | *Terminalia arjuna* (Roxb. ex DC.) Wight and Arn. | Arjun | | √ | | √ | √ | |
| 242. | *Terminalia bellirica* (Gaertn.) Roxb. | Beheda | √ | | | | | √ |
| 243. | *Thespesia populnea* (L.) Soland. ex Corr. | Paras pimpal | √ | √ | √ | | | |
| 244. | *Thuja plicata* ex D. Don | Vidya | | | | | | |
| 245. | *Tinospora cordifolia* (Willd.) Miers ex Hook. *f.* and Thoms | Gudvel | | √ | √ | √ | √ | |
| 246. | *Tribulus terrestris* L. | Gokhru; Gokshura mool | √ | | | | | |
| 247. | *Tricholepis indica* (Willd.) Almeida | Dahan | | √ | √ | √ | √ | |
| 248. | *Trichosanthes cucumerina* L. | Jangli padval | | | | | √ | |
| 249. | *Trichosanthes tricuspidata* Lour. | Mukul | | | | | √ | |
| 250. | *Tricticum aestivum* L. | Gahu | | | | | | |
| 251. | *Tridax procumbens* L. | Dagdi pala | | √ | | | √ | |
| 252. | *Vigna mungo* (L.) Hepper | Udid | | | | | | |
| 253. | *Vigna radiata* (L.) R. Wilczek | Sona mug | | | | | | |
| 254. | *Vigna trilobata* (L.) Verdc. | Mung | | | | | | |

| Sl.No. | Botanical Name | Local Name | Fruits | Leaves | Flowers | Stem | Roots | Seeds |
|---|---|---|---|---|---|---|---|---|
| 255. | *Vitex negundo* L. | Nirgundi | | √ | | | | |
| 256. | *Vitis vinifera* L. | Draksha | | | | | √ | |
| 257. | *Withania somnifera* (L.) Dunal | Ashwagandha | | √ | | | √ | |
| 258. | *Woodfordia fruticosa* (L.) Kurz. | Dhoneyti | | | √ | √ | | |
| 259. | *Wrightia tinctoria* R.Br. | Pandhra kuda | | √ | | √ | √ | |
| 260. | *Xanthium indicum* Koen. | Sankeshwar | | | | | √ | √ |
| 261. | *Zea mays* L. | Maka, Makki, Makai | | | | | | |
| 262. | *Zingiber officinale* Rosc. | Adrak | | | | | √ | |
| 263. | *Zizyphus mauritiana* Lam. | Bor | √ | √ | | √ | √ | |

juice (01 species). In case of monocotyledons, only 11 species are found useful for miscellaneous purposes (Table 6). Thus majority of dicotyledonous species (253 species) find 44 use-reports to sustain the life local people. Some trustees of temples or sacred yards maintain gardens. As such, they serve secondarily as centres for aesthetic beauty and recreation as the visitors or devotees are of all ages.

**Table 6: Plant Parts Used for Miscellaneous Purposes**

| *Sl. No.* | *Botanical Name* | *Bark* | *Fruits* | *Lea-ves* | *Flo-wers* | *Stem* | *Roots* | *Seeds* |
|---|---|---|---|---|---|---|---|---|
| 1. | *Abelmoschus esculentus* (L.) Moench. | | √ | | | | | |
| 2. | *Abelmoschus manihot* (L.) Medik. | | | | | √ | | |
| 3. | *Abitulon indicum* (L.) | | | | | | | |
| 4. | *Abrus precatorius* L. | | | | | | | √ |
| 5. | *Acacia chundra* (Roxb. ex Rottle.) Willd. | | | | | √ | | |
| 6. | *Acacia ferruginea* DC. | √ | | | | | | |
| 7. | *Acacia nilotica* (L.) Del. | | | | | | | |
| 8. | *Acalypha indica* L. | | | | | | | |
| 9. | *Achyranthes aspera* L. | | | | | √ | | |
| 10. | *Acorus calamus* L. | | | √ | | √ | √ | |
| 11. | *Adansonia digitata* L. | | | | | | | |
| 12. | *Aegle marmelos* (L.) Corr. | | | | | | | |
| 13. | *Ailanthus excelsa* Roxb. | | | | | | | |
| 14. | *Albizia lebbeck* (L.) Willd. | | | | | | | |
| 15. | *Allium cepa* L. | | | √ | | | √ | |
| 16. | *Allium sativum* L. | | | | | | | |
| 17. | *Aloe vera* L. | | | | | | | |
| 18. | *Alstonia scholaris* R.Br. | | | | | | | |
| 19. | *Amaranthus hybridus* L | | | | | | | |
| 20. | *Amaranthus spinosus* L. | | | √ | | | | |
| 21. | *Ammannia baccifera* L. | | | | | | | |
| 22. | *Andrographis echioides* (L.) Nees. | | | | | | | |
| 23. | *Anethum graveolens* L. | | | √ | | | | |
| 24. | *Anisomeles malabarica* R. Br. | | | | | | | |
| 25. | *Annona reticulata* L. | | | | | | | |
| 26. | *Annona squamosa* L. | | | | | | | |
| 27. | *Anogeissus latifolia* Wall. | | | | | | | |
| 28. | *Apium graveolens* L. | | | | | | | |

| Sl. No. | Botanical Name | Bark | Fruits | Lea-ves | Flo-wers | Stem | Roots | Seeds |
|---|---|---|---|---|---|---|---|---|
| 29. | *Arachis hypogaea* L. | | | | | | | |
| 30. | *Argemone mexicana* L. | | | | | | | |
| 31. | *Argyreia nervosa* (Burm. *f.*) Bojer | | | | | | | |
| 32. | *Aristolochia bracteolata* Lam. | | | | | | | |
| 33. | *Asparagus racemosus* Willd | | | | | | | |
| 34. | *Azadirachta indica* A. Juss. | | | | | | | |
| 35. | *Bacopa monnieri* Penn. | | | | | | | |
| 36. | *Balanitis aegyptiaca* (L.) Delile | | | | | | | |
| 37. | *Baliospermum raziana* Keshav *et.* Yog | | | | | √ | | |
| 38. | *Bambusa arundinacea* (Retz.) Willd. | | | | | √ | | |
| 39. | *Barleria lawii* L. | | | | | | | |
| 40. | *Barleria prionitis* L. | | | | | | | |
| 41. | *Bauhinia racemosa* Lamk. | | | √ | | | | |
| 42. | *Bauhinia tomentosa* L. | | | | | | | |
| 43. | *Bauhinia variegata* L. | | | | | √ | | |
| 44. | *Benincasa hispida* (Thunb.) Cogn | | √ | | | | | |
| 45. | *Boerhavia repens* L. var. *diffusa* (L.) Hook *f.* | | | | | | | |
| 46. | *Bombax ceiba* L. | | | | √ | | | |
| 47. | *Borassus flabellifer* L. | Juice | | | | | | |
| 48. | *Boswellia serrata* Roxb. | | | | | | | |
| 49. | *Bougainvillea spectabilis* Willd. | | | | | | | |
| 50. | *Bridelia retusa* (L.) Spreng. | | | | | | | |
| 51. | *Buchanania lanzan* Spreng. | | | | | | | |
| 52. | *Butea monosperma* (Lam.) Taub. | | | | | | | |
| 53. | *Caesalpinia bonduc* (L.) Roxb. | | | | | | | |
| 54. | *Caesalpinia pulcherrima* (L.) Sw. | | | | | | | |
| 55. | *Cajanus cajan* (L.) Millsp. | | | | | | | |
| 56. | *Calotropis gigantea* (L.) R. Br | | | | | | | |
| 57. | *Calotropis procera* (Ait.) R. Br. | | | | | | | |
| 58. | *Canna indica* L. | | | | | | | |
| 59. | *Cardiospermum halicacabum* L. | | | | | | | |
| 60. | *Carica papaya* L. | | | | | | | |
| 61. | *Cascabella thevetia* (L.) Lippold | | √ | | | | | |
| 62. | *Cassia absus* L. | | | | | | | |
| 63. | *Cassia auriculata* L | | | | | | | |

| Sl. No. | Botanical Name | Bark | Fruits | Lea-ves | Flo-wers | Stem | Roots | Seeds |
|---|---|---|---|---|---|---|---|---|
| 64. | *Cassia fistula* L. | | | | | | | |
| 65. | *Cassia occidentalis* L. | | | | | | | |
| 66. | *Cassia sophera* L. | | | √ | | | | |
| 67. | *Cassia tora* L. | | | | | | | |
| 68. | *Catharanthus roseus* (L.) G.Don | | | √ | | | | |
| 69. | *Celastrus paniculata* Willd. | | | √ | | √ | | √ |
| 70. | *Celosia argentea* L. | | | | | | | |
| 71. | *Centella asiatica* L. | | | √ | | | | |
| 72. | *Chlorophytum borivilianum* Sant. and Fernand. | | | | | | | |
| 73. | *Chrysanthemum indicum* L. | | | | | | | |
| 74. | *Cicer arietinum* L. | | | √ | | | | |
| 75. | *Cissampelos pareira* L. | | | √ | | | | |
| 76. | *Cissus quadrangularis* L | | | | | | | |
| 77. | *Citrullus colocynthis* Sch. | | | | | | | |
| 78. | *Citrullus lanatus* (Thunb.) Matsura and Nakai | | | | | | | |
| 79. | *Citrus aurantifolia* (Christm. and Panz.) Sw. | | | | | | | √ |
| 80. | *Citrus aurantium* L. | | | | | | | |
| 81. | *Cleome gynandra* L | | | | | | | |
| 82. | *Cleome viscosa* L | | | | | | | |
| 83. | *Clitoria biflora* Dalz. | | | √ | | | | |
| 84. | *Clitoria ternatea* L. | | | | | | | |
| 85. | *Coccinia grandis* (L.) Voigt | | | | | | | |
| 86. | *Cocculus hirsutus* (L.) Theob. | | √ | | | | | |
| 87. | *Cocos nucifera* L. | | | √ | | | | |
| 88. | *Corchorus capsularis* L. | | | | | | | |
| 89. | *Cordia dichotoma* Forst. *f.* | | | | | | | |
| 90. | *Coriandrum sativum* L. | | √ | | | | | |
| 91. | *Costus specious* (Koen.) J. E. Sm. | | | | | | | |
| 92. | *Crataeva adansonii* DC. | | | | | | | |
| 93. | *Crinum asiaticum* L. | | | | | | | |
| 94. | *Cucumis melo* L. | | | | | | | |
| 95. | *Cucumis sativus* L. | | | | | | | |
| 96. | *Cucurbita maxima* Duch. ex Lam. | | | | | | | |
| 97. | *Curculigo orchioides* Gaertn. | | | | | | | |

| Sl. No. | Botanical Name | Bark | Fruits | Lea-ves | Flo-wers | Stem | Roots | Seeds |
|---|---|---|---|---|---|---|---|---|
| 98. | *Curcuma amada* Roxb. | | | | | | | |
| 99. | *Curcuma inodora* Blatt. | | | | | | | |
| 100. | *Curcuma longa* L. | | | | | √ | | |
| 101. | *Cymbopogon citratus* (DC.) Stapf | | | | | | | |
| 102. | *Cymbopogon martinii* (Roxb.) Wats. | | | √ | | | | |
| 103. | *Cynodon dactylon* (L.) Pers. | | | | | | | |
| 104. | *Cyperus rotundus* L | | | | | | | |
| 105. | *Dalbergia sissoo* Roxb. ex DC. | | | | | | | |
| 106. | *Datura innoxia* Mill. | | | | | | | |
| 107. | *Datura metel* L. | | | | | | | |
| 108. | *Daucus carota* L. | | | | | | | |
| 109. | *Delonix regia* (Boj. ex Hook.) Raf. | | | | | | √ | |
| 110. | *Desmodium gangeticum* (L.) DC. | | | | | | | |
| 111. | *Diospyros melanoxylon* Roxb | | | | | | | |
| 112. | *Diospyros montana* Roxb | | | | | | | |
| 113. | *Diplocyclos palmatus* (L.) Jeffrey | | | | | | | |
| 114. | *Dolichandrone falcata* (Wall.ex DC) Seem. | | | | | | | |
| 115. | *Drimia indica* (Roxb.) Jessop | | | | | | | |
| 116. | *Echinops echinatus* Roxb. | | | | | | | |
| 117. | *Eclipta prostrata* (L.) L. | | | | | | | |
| 118. | *Elaeocarpus sphaericus* Gaertn. | | | | | | | |
| 119. | *Eleusine coracana* (L.) Gaertn. | | | | | | | |
| 120. | *Emblica officinalis* Gaertn. | | | | | | | |
| 121. | *Enicostema axillare* (Lam.) Raynal | | | | | | | |
| 122. | *Euphorbia hirta* L. | | | | | | | |
| 123. | *Evolvulus alsinoides* (L.) L. | | | | | | | |
| 124. | *Ficus benghalensis* L. | | | | | | | |
| 125. | *Ficus racemosa* L. | | | | | | | |
| 126. | *Ficus religiosa* L. | | | | | | | |
| 127. | *Ficus virens* Ait. | | | | | | | |
| 128. | *Flacourtia indica* (Burm. *f.*) Merr. | | | | | | | |
| 129. | *Garnga pinnata* Roxb. | | | | | | | |
| 130. | *Gloriosa superba* L. | | | | | | | |
| 131. | *Gmelina arborea* Roxb. | | | | | | | |
| 132. | *Gossypium arboreum* L. | | | | | | | |

| *Sl. No.* | *Botanical Name* | *Bark* | *Fruits* | *Lea-ves* | *Flo-wers* | *Stem* | *Roots* | *Seeds* |
|---|---|---|---|---|---|---|---|---|
| 133. | *Gossypium herbaceum* L. | | | | | | | |
| 134. | *Haldina cordifolia* (Roxb.) Ridsd. | | | √ | | √ | | |
| 135. | *Hardwickia binata* Roxb. | | | | | | | |
| 136. | *Helicteres isora* L. | | | | | √ | | |
| 137. | *Hemidesmus indicus* (L.) Schult. | √ | | | | | | |
| 138. | *Hibiscus rosa-sinensis* L. | | | | | | | |
| 139. | *Hibiscus sabdariffa* L. | | | | | | | |
| 140. | *Holarrhena pubescens* (Buch.-Ham.) Wall. ex G. Don | | | | | | | |
| 141. | *Holoptelea integrifolia* Planch. | | | | | | | |
| 142. | *Hymenodictyon ovixense* (Roxb.) Mabberly | | | | | | | √ |
| 143. | *Ipomoea illustris* (Clarke) Prain. | | | | | | | |
| 144. | *Ipomoea nil* (L.) Roth | | | | | | | |
| 145. | *Ixora coccinea* L. | | √ | √ | √ | √ | √ | √ |
| 146. | *Jasminum officinale* L. | | | | | | | |
| 147. | *Jasminum sambac* (L.) Ait. | | | | √ | | | |
| 148. | *Jatropha curcas* L. | | | | | | | |
| 149. | *Justicia adhatoda* L. | | | | | | | |
| 150. | *Kalanchoe pinnata* (Lam.) Pers. | | | | | | | |
| 151. | *Kohautia aspera* (Heyne ex Roth) Brem. | | | | | | | |
| 152. | *Lagenaria siceraria* (Mol.) Standl. | | | | | | | |
| 153. | *Lagerstroemia microcarpa* Wight | | √ | | | | | |
| 154. | *Lantana camara* L. | | | | | | | |
| 155. | *Lawsonia inermis* L. | | | | | | | |
| 156. | *Leucas cephalotes* (Roth) Spr. | | | | | | | |
| 157. | *Limonia acidissima* L. | | | | | | | |
| 158. | *Luffa acutangula* (L.) Roxb. | | | | | | | |
| 159. | *Madhuca longifolia* (Koen.) Mac Bride | | √ | | | | | |
| 160. | *Mallotus philippensis* (Lam.) Muell-Arg. | √ | | √ | | √ | | |
| 161. | *Mangifera indica* L. | | | | | | | |
| 162. | *Martynia annua* L. | | | | | | | |
| 163. | *Melia azedarach* L. | | | | | | | |
| 164. | *Melilotus alba* Medik. ex Desr. | | | | | | | |
| 165. | *Merremia dissecta* (Jacq.) Hallier *f.* | | √ | | | | | |

| Sl. No. | Botanical Name | Bark | Fruits | Leaves | Flowers | Stem | Roots | Seeds |
|---|---|---|---|---|---|---|---|---|
| 166. | *Michelia champaca* L. | | √ | √ | √ | √ | √ | √ |
| 167. | *Mimosa hamata* Willed. | | | | | | | |
| 168. | *Mimusops elengi* L. | | | | | | | |
| 169. | *Mirabilis jalapa* L. | | √ | | | | | |
| 170. | *Mitragyna parvifolia* (Roxb.) Korth. | | | | | | | |
| 171. | *Momordica dioica* Roxb. ex Willd. | | | | | | √ | |
| 172. | *Moringa oleifera* Lam. | | √ | | | | | |
| 173. | *Mucuna pruriens* (L.) DC. | | | | | | | √ |
| 174. | *Mukia maderaspatana* L. | | | | | √ | | |
| 175. | *Musa paradisiaca* L. | | | | | | | |
| 176. | *Nelumbo nucifera* Gaertn. | | | | | | √ | |
| 177. | *Nerium indicum* Mill. | | | | √ | | | |
| 178. | *Nyctanthes arbortristis* L. | | | | | | | |
| 179. | *Nymphaea nouchali* Burm.*f.* | | | | √ | | | |
| 180. | *Nymphaea pubescens* Willd | | | | | | | |
| 181. | *Ocimum basilicum* L. | | | | | | | |
| 182. | *Ocimum gratissimum* L. | | | | | | | |
| 183. | *Ocimum tenuiflorum L.* | | | | | | | |
| 184. | *Opuntia elatior* Mill. | | | | | | | |
| 185. | *Oroxylum indicum* (L.) Vent. | | | | | | | |
| 186. | *Oryza sativa* L. | | | | | | | |
| 187. | *Oxalis corniculata* L. | | | | | | | |
| 188. | *Pandanus odoratissimus* L. *f.* | | √ | √ | √ | √ | √ | √ |
| 189. | *Passiflora edulis* Sims | | | | | | | |
| 190. | *Pedalium murex* L. | | | | √ | | | |
| 191. | *Pennisetum americanum* L. | | | | | | | |
| 192. | *Pergularia daemia* (Forssk.) Chiov. | | | | | | | |
| 193. | *Phoenix dactylifera* (L) Roxb. | | | | | | | |
| 194. | *Phyla nodiflora* (L.) Greene | | | | | | | |
| 195. | *Phyllanthus acidus* (L.) K. Skeels | | | | | | | |
| 196. | *Phyllanthus amarus* L. | | | | | | | |
| 197. | *Physalis micrantha* L. | | | | | | | |
| 198. | *Piper betle* L. | | √ | | | | | |
| 199. | *Plumbago zeylanica* L. | | | | | | | |
| 200. | *Plumeria rubra* L. forma *acuminata* (Ait.) Santapaul and Irani ex Shah | | | | | | | |

| *Sl. No.* | *Botanical Name* | *Bark* | *Fruits* | *Lea-ves* | *Flo-wers* | *Stem* | *Roots* | *Seeds* |
|---|---|---|---|---|---|---|---|---|
| 201. | *Polyalthia longifolia* Sonn. | | | | | | | |
| 202. | *Pongamia pinnata* (L.) Pierre | | | | | | | |
| 203. | *Prosopis cineraria* (L.) Druce | | | | | | | |
| 204. | *Psidium guajava* L. | | | | | | | |
| 205. | *Pterocarpus marsupium* Roxb. | | | | | | | |
| 206. | *Pueraria tuberosa* (Roxb. ex Willd.) DC. | | | | | | | |
| 207. | *Punica granatum* L. | | | | | | √ | |
| 208. | *Quisqualis indica* L. | | √ | | | | | |
| 209. | *Raphanus sativus* L. | | | | | | | |
| 210. | *Ricinus communis* L. | | | √ | | | | |
| 211. | *Rosa damascena* Mill | | | | | | | |
| 212. | *Saccharum officinarum* L. | | | | √ | | | |
| 213. | *Salvadora persica* L. | | | | | | | |
| 214. | *Santalum album* L. | | | | | | | |
| 215. | *Sapindus emarginatus* Vahl. | | | | | | | |
| 216. | *Saraca asoca* (Roxb.) de Wilde | | √ | | | | | |
| 217. | *Schleichera oleosa* (Lour.) Oken | | | | | | | |
| 218. | *Securinega virosa* (Roxb. ex Willd.) Baill. | | | | | | | √ |
| 219. | *Semecarpus anacardium* L. | | | | | | | |
| 220. | *Sesamum orientale* L. | | | | | | | |
| 221. | *Sesbania grandiflora* (L.) Poir. | | | | | | | √ |
| 222. | *Sesbania sesban* (L.) Merr. | | | | √ | | | |
| 223. | *Setaria italica* (L.) P. Beauv. | | | | | | | |
| 224. | *Sida acuta* Burm. *f.* | | | | | | | |
| 225. | *Sida cordata* (Burm.*f.*) Borss. | | | | | | | |
| 226. | *Sida spinosa* L. | | | | | | | |
| 227. | *Solanum anguivi* Lam. | | | | | √ | | |
| 228. | *Solanum melongena* L. | | √ | | | | | |
| 229. | *Solanum nigrum* L. | | √ | | | | | |
| 230. | *Solanum virginianum* L. | | | | | | | |
| 231. | *Sorghum vulgare* Pers. | | | | | | | |
| 232. | *Spilanthus calva* DC. | | | | | | | |
| 233. | *Spondias pinnata* (L. *f.*) Kurz. | | | | | | | |
| 234. | *Sterculia urens* Roxb. | | √ | | | | | |
| 235. | *Syzygium cumini* (L.) Skeels | | | | | | | √ |
| 236. | *Tabernaemontana divaricata* (L.) R. Br. | | | | | | | |

| Sl. No. | Botanical Name | Bark | Fruits | Leaves | Flowers | Stem | Roots | Seeds |
|---|---|---|---|---|---|---|---|---|
| 237. | *Tagetes erecta* L. | | | | | | | |
| 238. | *Tamarindus indica* L. | | | | | | | |
| 239. | *Tectona grandis* L. | | | | | | | |
| 240. | *Tephrosia purpurea* Pers. | | | | | | | |
| 241. | *Terminalia arjuna* (Roxb. ex DC.) Wight and Arn. | | | | | | | |
| 242. | *Terminalia bellirica* (Gaertn.) Roxb. | | | | | | | |
| 243. | *Thespesia populnea* (L.) Soland. ex Corr. | | | | | | | |
| 244. | *Thuja plicata* ex D. Don | | | | | | | |
| 245. | *Tinospora cordifolia* (Willd.) Miers ex Hook. *f.* and Thoms | | | | | | | |
| 246. | *Tribulus terrestris* L. | | | | | | | |
| 247. | *Tricholepis indica* (Willd.) Almeida | | | | | | | |
| 248. | *Trichosanthes cucumerina* L. | | | | | | | |
| 249. | *Trichosanthes tricuspidata* Lour. | | | | | | | |
| 250. | *Tricticum aestivum* L. | | | | | | | |
| 251. | *Tridax procumbens* L. | | | | | | | |
| 252. | *Vigna mungo* (L.) Hepper | | | | | | | |
| 253. | *Vigna radiata* (L.) R. Wilczek | | | | | | | |
| 254. | *Vigna trilobata* (L.) Verdc. | | | | | | | |
| 255. | *Vitex negundo* L. | | | | | | | |
| 256. | *Vitis vinifera* L. | | | | | | | |
| 257. | *Withania somnifera* (L.) Dunal | | | | | | | |
| 258. | *Woodfordia fruticosa* (L.) Kurz. | | | | | | | |
| 259. | *Wrightia tinctoria* R.Br. | | | | | | | |
| 260. | *Xanthium indicum* Koen. | | | | | | | |
| 261. | *Zea mays* L. | | | | | | | |
| 262. | *Zingiber officinale* Rosc. | | | | | | | |
| 263. | *Zizyphus mauritiana* Lam. | | | | | | √ | |

## Indigenous People as Conservationist

Plants, the greatest gift of Mother Nature to mankind, have been regarded with special reverence. Plant or tree worship is closely allied with the sacred and even superstitious reverence from time immemorial. Plants are also believed as the abodes of local deities. There are customs or taboos to protect and conserve plant species, certain pockets of forests or even entire forests. Such sacred groves are observed throughout over country (*cf.* Vartak and Gadgil, 1981, 1990). Since this pioneering investigation on sacred groves or shrines, there has been mushrooming growth in literature published pertaining to socio-culture, conceptual, historical, functional attributes of sacred groves, *etc.* It is undeniable that plants which are used as temple offerings and associated with religio-cultural beliefs are commonly grown or wild ones are protected in the courtyards of the temples (whether in primitive form or well built), shrines or groves. Interlinking religion with nature help conserve the latter. This is taught to the indigenous society traditionally. The groves, shrines and temple yards maintain significant level of biodiversity useful for various purposes as stated earlier. These areas naturally act as reservoirs of biological diversity and contribute directly towards its conservation and sustainable utilization. These reservoirs certainly minimize genetic erosion of the biodiversity in the said region, and even save from erosion of traditional knowledge. Khandesh region is studded with many religious and temple courtyards (Appendix I) and as such provided places for plantations.

## Study of Ancient Indian Literature

Certain plant species regarded sacred in this part of Maharashtra are also noted so in Vedic texts, *e.g. Ficus benghalensis* L., *Ficus religiosa* L., *Pongamia pinnata* (L.) Pierre, *Butea monosperma* (Lam.) Taub., *Nelumbo nucifera* Gaerth, *Cymbopogon citratus* (DC.) Stapf, *Bambusa arundinacea* (Retz.)Willd., and few others. The other Vedas and ancient scripts and epics are replete with references to sacred plants. Indians during Vedic, even post-Vedic and modern periods have regarded all of nature, in some sense, divine. Plants, animals, earth and other planets are united and held sacred. According to Atharveda, medicinal plants are regarded goddesses. If analyze with our local faith and belief with those plant species, we may find parallel or continuity in faith and belief even in modern time. Comparative and cross-cultural comparison of sacred and religious plants still await for further studies including ancient Sanskrit scripts (Patil, 2020).

## Economic Significance

It is a common scene at some famous and notable temples that articles or plant parts are sold outside the yards of temples (Photographs:30-39). The common ones are: coconuts, floral garlands, rosaries, essence sticks, turmeric powder, flowers and the leaves of some religious useful plants. The present investigators recorded some common ones, although sometimes not a part of the temple courtyards as the following: (i) Leaves: *Aegle marmelos* (L.) Corr., *Bauhinia racemosa* Lamk., *Calotropis gigantea* (L.) R. Br, *Ocimum tenuiflorum* L., *Cynodon dactylon* (L.) Pers, *Mangifera indica* L., *Piper betle* L., *Plumeria rubra* L, *Polyalthia longifolia* (Sonner.) Thw. (substitute for mango leaves), *Sesbania grandiflora* (L.) Poir, *Jasminum officinale* L., *Jasminum sambac* (L.) Ait. (ii) Entire plants: *Achyranthes aspera* L., *Celosia argentea* L., *Cyperus rotundus* L, *Musa*

**Photograph 30: Jaswand (*Hibiscus rosa-sinensis*), Bel (*Aegle marmelos*) and Durva (*Cynodon dactylon*) in the worship of Lord Ganesha at Padmalaya Temple (Jalgaon district).**

**Photograph 31: A person during steam bath of the leaves of Nirgudi (*Vitex negundo*) while treating paralysis.**

**Photograph 32: Local Vendors of Manudevi (Jalgaon district) selling various religious botanicals.**

**Photograph 33: Selling Durva (*Cynodon dactylon*), Zendu (*Tagatus erectus*) and Jaswand (*Hidiscus rosa-sinensis*) at Muktai Mandir, Hingane, Muktainagar (Jalgaon district).**

**Photograph 34: Padmalaya (Jalgaon district): Selling religious botanicals.**

**Photograph 35: Vaidu (Medicine woman) Naginabai at Patanadevi (Jalgaon district).**

**Photograph 36 : Unapdev Yearly Fair (Jalgaon district): selling religious various botanicals.**

**Photograph 37: Dudhi-bhopala (*Lagenaria siceraria*) vendor at Toranmal roadside (Nandurbar district).**

**Photograph 38: Herbal vendor at Toranmal (Nandurbar district).**

**Photograph 39: Avala (*Emblica officinalis*): vendor at Toranmal (Nandurbar district).**

*paradisiaca* L, *Saccharum officinarum* L. *etc.* (iii) Flowers: *Hibiscus rosa-sinensis* L., *Polianthes tuberosa* L., *Chrysanthemum indicum* L., *Tagetes erecta* L, *Tagetes patula* L. *Datura innoxia* Mill., *Datura metel* L., *Crossandra infundibuliformis* (L.) Nees., *Ipomoea quamoclit* L., *Jasminum auriculatum* Vahl., *Michelia champaca* L., *Nyctanthes arbortristis* L., *Clitoria ternatea* L., *Canna orchioides* Bailey, *Nerium indicum* Mill., *Rosa damascena* Mill., *etc.* (iv) Fruits: *Aegle marmelos* (L.) Corr., *Annona squamosa* L., *Citrus aurantifolia* (Christm.) Sw., *Citrus aurantium* L., *Capsicum annuum* L., *Semecarpus anacardium* L.*f*., *Cocos nucifera* L., *Emblica officinalis* Gaerth, *etc. Punica granatum* L. *etc.* (v) Garlands: *Tagetes erecta* L., *Tagetes patula* L., *Polianthes tuberosa* L., *Nyctanthes arbortristis* L., *Jasminum grandiflorum* L., *Jasminum sambac* (L.) Ait., *Rosa damascena* Mill., *Chrysanthemum indicum* L. *etc.* (vi) Rosaries: (i) Stem of *Ocimum tenuiflorum* L. (called Tulsimala), (ii) *Elaeocarpus sphaericus* Gaertn. (Syn. *E.ganitrus* Roxb. ex G.Don) (called Rudraksh mala).

All these plants parts have special demands on some festivals, ceremonies, pilgrims, worships on particular day of a week or month or entire month (like Shrvana). Many festivals (tribal and non-tribal), yatras, rites, pilgrims, deities are notable (Tables 2-5 and 8). These afford chances to sell botanicals for religious purposes. They are marketed on large scale. Many of them are cultivated ones and collected from private home gardens. Floriculture is also practiced by few farmers in the region. Some species are gathered from wild sources. The species *viz. Aegle marmelos* (L.) Corr. and *Bauhinia racemosa* Lamk. are stripped off extensively for their foliage on large scale and therefore these trees are always rendered to precarious condition. Some other cultigens or wild sources are sold as noted in the enumeration.

## Model Gardens for Religious and Sacred Plants

After scrutiny of religious and sacred plants demanded or needed in the region for various rites, rituals, ceremonies, festivals, pilgrims, temples and market availability, *etc.*, the present investigators thought worthwhile to put an ideas of garden of these most demanded plant species. Such garden can be maintained by the trusts of temples or individually by the interested ones to provide these plants or their parts as and when demanded by the various societies. This may help uplift the economy in the region and may have access in reasonable period for such plant sources. The categories (i) to (v) and plant species mentioned above under economic significance above are suggested to have in their gardens. Such a gardener, however, should first know the various social activities and special demand in his area. The number of species may be selected depending upon the space available to the garden planner.

## Temple Architecture

Amalakah is Sanskrit name for the fruit of Phyllanthus emblica Linn. (Syn. Emblica officinalis Gaertn.), the Indian Gooseberry. A vertically grooved, circular structure crowning the tower of a temple design in India is called Amala.

It simulates fruits of Amalakah (*cf.*) (Singh and Sharma, 2008; Singh, 2014). This design is also found atop the Kapileshwar Temple at Mudawad Tq. Amalner, District Jalgaon, (M.S.) (Photograph 26). This tree species is deeply rooted in Indian bioculture w.r.t. myths, legends, beliefs, rites, rituals, deities, worships, architecture, *etc.* (Patil, 2020). Temples in India are usually associated with the tree called 'Pipal' or 'Pimpal' (*Ficus religiosa* L.) as in this photograph.

## Ethnoeconomy

The diversity and conservation of sacred plants play a significant role in the sustainable development of rural and tribal population. Successful co-existence of the plant and man is tricky situation. Although man has fulfilled basic need through the plants, which they find in their surrounding wealth they, in turn, jeopardize the health of their wealth. Biodiversity is a part of our daily life and livelihood and constitutes the resources upon which families, communities, nations and future generations depend. Many minor forest products (MFP's) are important not only in rural and tribal economics, but also in the national economy. Mamkind depends on MFP's for food energy, construction materials, medicine, *etc.* These resources have the critical character of being renewable, hence with the proper management, they can be used sustainably ensuring their use for future generations.

In Jalgaon, Dhule and Nandurbar districts as a majority of the tribal and rural house holders are both marginal and small cultivators. The pressure on land is heavy and such householders are not in a position to fulfil their needs. The average size of the land holding is hardly 2 ha. Only few tribals and rurals have land size larger than 2 ha. Moreover, due to several factors such as poor soil fertility, uncertain monsoon, lack of irrigation facilities, poverty illiteracy, the villagers get poor yield from agriculture.

So large number of tribal and rural householders have resorted to subsidiary occupation for meeting their minimum requirements. The main source of cash income for majority of tribal and rurals of this area is from collection and sale of firewood and other minor forest products for food, energy, construction materials, medicine, honey and fodder, *etc.* Many of the MFP's are being used by aboriginals for their day to day needs and many of them are income generative sources. It was observed that consumption of Bamboo [*Bambusa arundinacea* (Retz.)Willd.] is very high being used for making many items like basckets, rope holders, grain covers, shelves, poultry hatchery roofing of house, pig stay livestock room, fencings *etc.* There are several other items which tribal peoples use in their day to day and are made from forest product like grain crusher from the wood of *Acacia chundra* (Roxb. ex Rottle) Willd. Lamp stands are made out of *Terminalia arjuna* (Roxb. ex DC.), mats are woven from the leaf blades of *Borassus flabellifer* L. (Shindhi). The ropes are made from *Hibiscus cannabinas* L. (Ambadi), *Cyperus rotundus* L. and *Helicteres isora* L.(Murudsheng). They prepare pesticide for crops from the leaves of *Azadirachta indica* A. Juss. Most of the fallen branches and leaves of the forest species are being used as fodder. Sometimes they cut the branches for their daily needs. An average quantity of fuel wood used per day is 5 kg. (fallen branches) per house. Different MFP's and the product values are tabulated in Table 6.

They also collect gum, honey, leaves of Palas [*Butea monosperma* (Lam.) Taub.] to make dinner plates and bowls, Mahu [*Madhuca longifolia* (Koen.) Mac Bride] flowers, Mahu seeds, flowers, fruits like Bor (*Zizyphus jujuba* Mill.), Chinch (*Tamarindus indica* L.), Awala (*Emblica officinalis* Gaertn), Charoli (*Buchanania lanzan* Spreng.) have been revealed that those MFP's contribute fairly to total income of household to the tribals.

The sacred plants, their twigs, flowers, fruits and leaves also play important role in income generation of rural and tribal people during various local festivals like Diwali, Dashara, Sankrat, Navaratri, Gauri pooja, Ganapati festival (Tables 7-9 and 12). Many botanicals are essential for worship and sold in market during these days. Also at the time of Buddh Pornima, Jain festivals and at Pateti, various plants parts and flowers are used for sacred offerings, which are available in market. During the marriage ceremony, birth celebration, birth anniversary, rituals and rites, flowers, twigs, fruits are generally used. In villages and in sacred forests many deities and their temple are present. The fair called 'Jatra' are held during particular period of the year. Where marketing of honey, medicinal plant products, sticks, deity idol, flowers, fruits and twigs are sold as offering for deities. Wooden utensils are also sold. Most of the villagers, also urban a tribal people visit these fairs. They have very strong faith and belief for the forest situated deities called 'Kuldevata'. Some villagers are involved in this economic activity and help to improve their economy and livelihood. Sacred plant parts and their values in market have been given in Tables 2, 3 and 4. In tribal region, fair of Yahimoga Mata, Marriage festivals like 'Bhongarya' of Pawara tribe, many plant parts and medicinal plants are sold in tribal hamlets and nearby village. Religious botanicals are usually sold near temples or religious places or at the time of ceremonies, festivals, *etc.* (Ghate, 1998; Pawar and Patil, 2010).

MFP's and sacred plants supports economic interaction of the tribal and rurals with outside world. Though the tribals are inward oriented, on certain counts they are required to come into contact with the non tribal world through exchange, through the process of sale many MFP's and sacred plant products. In this activity, forests have an important role to play. Forests are sources for fuelwood, bamboo and MFPs. However, since the forest are degraded, their economy base has also shrunk. Hence depletion of forest is directly related with the weakening of economical base. Any thoughtful programme of tribal development ought to take this aspect in to proper perspective. The plant sources employed for local economy and briefly presented in the Flow Diagram-I depictinig sustainability in the region. Likewise, Photographs 30-39 informs about botanicals sold in religious places.

**Flow Diagram-I: Minor Forest Products**

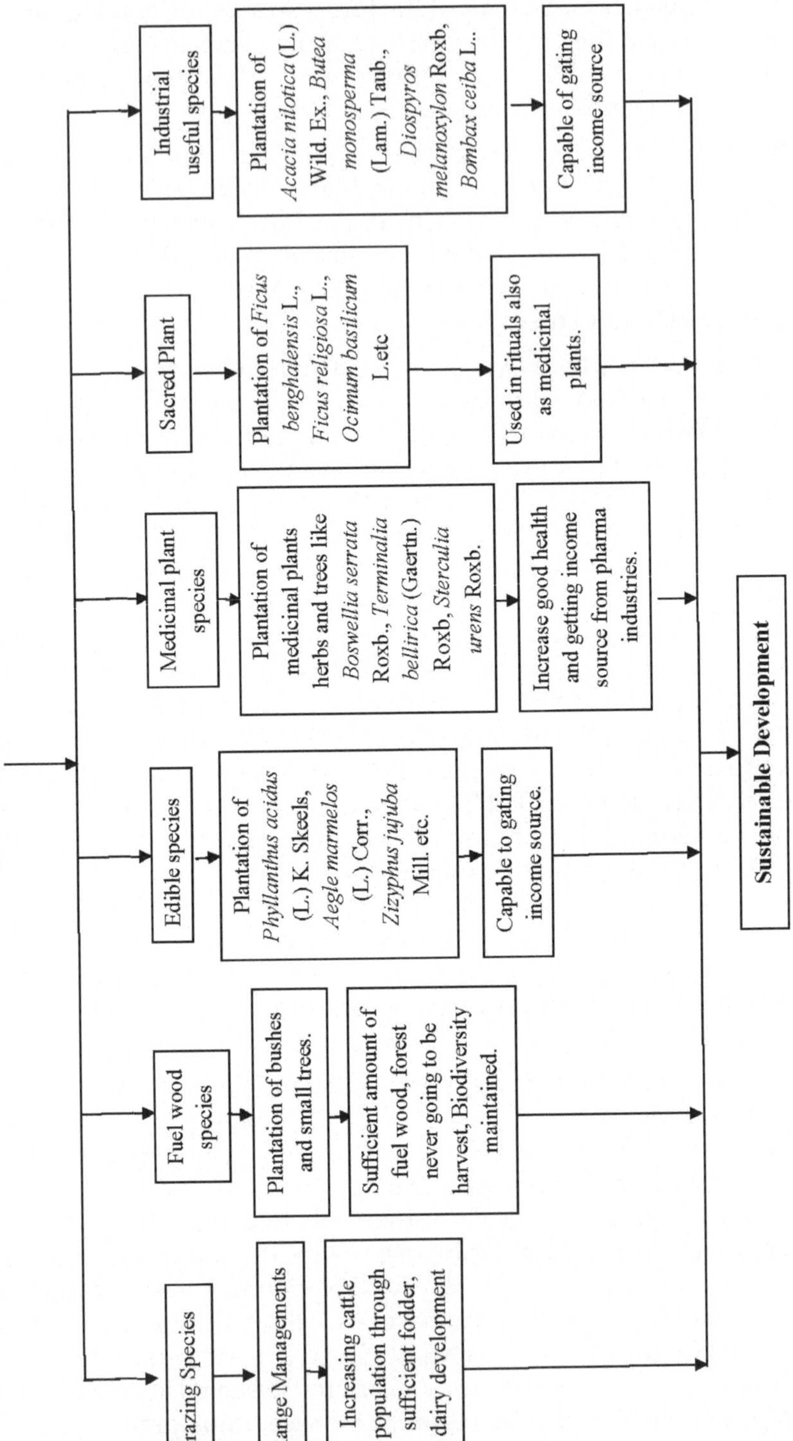

**Table 7: Plants Used for Sacred Purpose: Flowers**

| *Sl. No.* | *Botanical Name* | *Local/Commercial Name* | *Plant Part Use* | *Value (Rs./kg)* |
|---|---|---|---|---|
| 1. | *Clitoria ternatea* L. | Gokarna | Flower | 1 Rs./1 flower |
| 2. | *Hibiscus rosa-sinensis* L. | Jaswand | Flower | 2-3 Rs./each flower |
| 3. | *Jasminum officinale* L. | Chameli | Flower | 15-25 Rs./1 Festoon |
| 4. | *Jasminum sambac* (L.) Ait. | Mogara | Flower | 15-25 Rs./1 Festoon |
| 5. | *Michelia champaca* L. | Son chafa | Flower | 1-2 Rs./1 Flower |
| 6. | *Nelumbo nucifera* Gaertn. | Kamal | Flower | 10 Rs./1 bunch of 5 flowers |
| 7. | *Rosa damascena* Mill. | Gulab | Flower | 5-10 Rs./1 flower |
| 8. | *Tabernaemontana divaricata* (L.) R. Br. | Tagar | Flower | 10 Rs./basket with other worship ingradient. |
| 9. | *Tagetes erecta* L. | Zendu | Flower | 60 Rs./Kg. |

**Table 8: Plants Used for Sacred Purpose: Fruits**

| *Sl. No.* | *Botanical Name* | *Local/Commercial Name* | *Plant Part Use* | *Value (Rs./kg)* |
|---|---|---|---|---|
| 1. | *Abelmoschus esculentus* L. | Bhendi | Fruit | 25-30 Rs./Kg. |
| 2. | *Annona squamosa* L. | Sitaphal | Fruit | 55-70 Rs./Kg. |
| 3. | *Benincasa hispida* (Thunb.) Cogn. | Kohla | Fruit | 25-30 Rs./1 Fruit |
| 4. | *Carica payaya* L. | Papai | Fruit | 20-30 Rs./1 Fruit |
| 5. | *Citrus aurantium* L. | Santra | Fruit | 60-70 Rs./Kg. |
| 6. | *Citrus limon* (L.) Burm. | Limbu | Fruit | 2 Rs./1 fruit |
| 7. | *Coccinia grandis* (L.) Voigt | Tondli | Fruit | 15-20 Rs./Kg. |
| 8. | *Cocos nucifera* L. | Naral | Fruit | 15-20 Rs./1fruit |
| 9. | *Cucumis melo* L. | Kharbuz | Fruit | 15-20 Rs./Kg. |
| 10. | *Cucumis sativus* L. | Kakadi | Fruit | 10-20 Rs./Kg. |
| 11. | *Cucurbita maxima* Duch. ex Lam. | Bhopala | Fruit | 25-30 Rs./Kg. |
| 12. | *Emblica officinalis* Gaertn. | Aawala | Fruit | 80 Rs./Kg |
| 13. | *Mangifera indica* L. | Aamba | Fruit | 55-60 Rs./Kg |
| 14. | *Musa paradisiaca* L. | Keli | Fruit | 25-30 Rs./Dozen |
| 15. | *Psidium guajava* L. | Peru | Fruit | 25-40 Rs./Kg. |
| 16. | *Punica granatum* L. | Dalimb | Fruit | 40-60 Rs./Kg. |
| 17. | *Sapindus emarginatus* Vahl. | Ritha | Dried Fruits | 40 Rs./100gm |
| 18. | *Solanum melongena* L. *var.* melongena | Vangi | Fruit | 20-30 Rs./Kg |

| *Sl. No.* | *Botanical Name* | *Local/Commercial Name* | *Plant Part Use* | *Value (Rs./kg)* |
|---|---|---|---|---|
| 19. | *Terminalia belerica* (Gaertn.) Roxb. | Behada | Dried Fruits | 40 Rs./100gm |
| 20. | *Terminalia chebula* Retz. | Hirda | Dried Fruits | 40 Rs./100gm |
| 21. | *Zizyphus jujuba* Mill. | Bor | Fruit | 25-45 Rs./Kg. |

**Table 9: Plants Used for Sacred Purpose: Leaves**

| *Sl. No.* | *Botanical Name* | *Local/Commercial Name* | *Plant Part Use* | *Value (Rs./kg)* |
|---|---|---|---|---|
| 1. | *Aegle marmelos* (L.) Corr. | Bel | Leaves | 10-20 Rs./1 small basket. |
| 2. | *Cynodon dactylon* (L.) Pers. | Durva | Young grass needles | 5 Rs./bunch needles |
| 3. | *Musa paradisiaca* L. | Keli | Leaves | 10 Rs./10 Leaves |
| 4. | *Trigonella foenum-graecum* L. | Methi | Leaves and twig | 20 Rs./Kg. |

**Table 10: Plants Used for Sacred Purpose: Others**

| *Sl. No.* | *Botanical Name* | *Local/Commercial Name* | *Plant Part Use* | *Value (Rs./kg)* |
|---|---|---|---|---|
| 1. | *Arachis hypogaea* L. | Bhuimug | Nuts | 55-70 Rs./Kg. |
| 2. | *Santalum album* L. | Chandan | Sap wood | |
| 3. | *Sesamum orientale* L. | Til | Seeds | 160-190 Rs./Kg. |

## Conservation

India has a rich tradition of nature conservation as well as a vigorous official program of nature reserves developed over the last 40 years. Various indigenous, communities lived in harmony with nature and thus conserved biodiversity. In the course of time science and technology developed. Industries were established and expanded to meet the increasing demands of people. Various developmental activities, over increasing population and growth of infrastructure facilities have resulted in decline of biodiversity, habitat alteration, over exploitation, pollution and introduction of exotic species also threatened the global biological resources. Modernization and commercialization of agriculture in order to increase production are the cause of disappearing traditional knowledge among people. There is strong need to initiate people's participation, training to promotion of indigenous traditional knowledge and conserve biodiversity through traditional knowledge.

**Table 11: Plants Used for Medicinal Purpose**

| *Sl. No.* | *Botanical Name* | *Local/Commercial Name* | *Plant Part Use* | *Value (Rs./kg)* |
|---|---|---|---|---|
| 1. | *Acacia concinna* (Willd.) DC. | Shikekai | Dried pods | 55-70 Rs./Kg. |
| 2. | *Acorus calamus* L. | Vekhand, Vacha | Dride twig/ Powder | 10 Rs/100 gms. |
| 3. | *Aegle marmelos* (L.) Corr. | Bel | Gar | 120 Rs./Kg. |
| 4. | *Allium cepa* L. | Kanda | Rhizome | 15-40 Rs./Kg. |
| 5. | *Allium sativum* L. | Lasun | Rhizomes | 160-200 Rs./Kg. |
| 6. | *Aloe vera* L. | Korphad | Latex | 40-60 Rs./100 ml. |
| 7. | *Azadirachta indica* A. Juss. | Neem | Small twig | 5-10 Rs./Twig |
| 8. | *Borassus flabellifer* L. | Tad | Fruit Juice "Neera" | — |
| 9. | *Boswellia serrata* Roxb. | Salai, | BarkGum | 250 Rs./Kg—— |
| 10. | *Chlorophytum borivilianums* Santapau and Fernad. | Safed musali | Tubers | 1800 Rs./Kg |
| 11. | *Chlorozophora prostrate* Dalz. | Unhali | Roots | 150 Rs./Kg |
| 12. | *Citrullus colocynthis* (L.) Schrad. | Kadu indravan | Fruit | 160 Rs./Kg |
| 13. | *Crinum latifolium* L. | Jangali kanda | Roots | 100 Rs./Kg. |
| 14. | *Curcuma amada* Roxb. | Ambe Halad | Dried Rhizome Powder | 10 Rs./50 gms. |
| 15. | *Curcuma longa* L. | Halad | Powder | 120-160 Rs./Kg |
| 16. | *Cymbopogon citratus* (DC.) Stapf | Gavati Chaha | Leaves | 5 Rs./bunch |
| 17. | *Ensete superbum* (Roxb.) Cheesm. | Jangali keli | Fruit | 100 Rs./Kg |
| 18. | *Eulophia herbacea* Lindl. | Kukkut kand | Root and Tubers | 150 Rs./Kg |
| 19. | *Helicteres isora* L. | Murudsheng | Pods | 10 Rs./100 gms |
| 20. | *Madhuca longifolia* (Koen.) Mac Bride *var. longifolia* | Moha | Fruit Beverage | 5 Rs./1 glass |
| 21. | *Mentha arvensis* L. | Pudina | Leaves | 15-20 Rs./Small bunch |
| 22. | *Tamarindus indica* L. | Imli, Chinch | Dried pods. | 40-50 Rs./Kg. |
| 23. | *Tectona grandis* L. | Sag | Seeds | 1200 Rs./Kg. |
| 24. | *Terminalia arjuna* (Roxb.) Wight and Arn. | Arjun saal | Bark | 500 Rs./Kg. |

**Table 12: Minor Forest Products**

| Sl. No. | Plant Name | Plant Part | Use | Income Generate |
|---|---|---|---|---|
| 1. | *Butea monosperma* (Lam.) Taub. | Leaf | Making plates | 15Rs./100 leaves |
| 2. | *Diospyros melanoxylon* Roxb. | Leaf | Beedi preparation | 200 Rs/100 leaves |
| | | Fruit | Jam preparation | 10 Rs./kg. |
| 3. | *Cassia tora* L. | Leaves | Vegetable | 3-5 Rs./kg. |
| 4. | *Madhuca longifolia* (Koen.) Mac Bride | Flower | Alcoholic beverages and sweetener | 15 Rs./kg. |
| | | Seed | —— | 20 Rs./kg. |
| 5. | *Syzygium cumini* (L.) Skeels | Fruit | Edible fruit | 100 Rs./kg. |
| 6. | *Buchanania lanzan* Spreng. | Seeds | Edible fruit | 150 Rs./kg. |
| 7. | *Emblica officinalis* Gaertn. | Fruit | Candy, jam, jelly. | 40 Rs./kg. |
| 8. | *Zizyphus jujuba* Mill. | Fruit | Edible fruit | 30 Rs./kg. |
| 9. | *Tamarindus indica* L. | Fruit | Vegetable | 40 Rs./kg. |
| 10. | *Boswellia serrata* Roxb. | Gum | In sweets | 150 Rs./kg. |
| 11. | *Acacia nilotica* (L.) Wild. ex | Gum | In sweets | 100 Rs./kg. |
| 12. | *Sterculia urens* Roxb. | Gum | In sweets | 100 Rs./kg. |
| | | Fallen branches, twigs | Construction | 100-1200 Rs./kg. |
| | | Wood | Cooking | 10 Rs./kg. |

## Ancient Ethics of Conservation

The concept and beliefs of sacred trees and sacred forest is one of the best practices to conserve the natural resources. However, very little information is available on sacred trees and conservation of biodiversity in Khandesh region. People of Khandesh follow ancestral worship and animism. In the form of deity worship, with the central focus of worship on forest patches *e.g.* Satpuda region, near Dhadgaon. Akkalkuwa, Nandurbar, Navapur, patches of Satpuda formed northern side as Raver, Pal, Chopada, Yawal. Also Hasti range of forrest near Jamner and Ajanta, Pitalkhora near Chalisgaon tehsil, which signity sacred groves, Temple groves, sacred sites *etc.*, where number of sacred plants are worshiped. Important religious shrines are located beside river Bori, Tapi, Girna and its tributaries.

Trees have been regarded as most important part of life and their importance described in epics has great impact on mankind. They maintained equilibrium for the subsistence of life till the last century. But increasing demands of population and industrialization started to disturb this equilibrium. Natural resources live land, water, air and plants are considered as free and inexhaustible resources. The basic element of nature in the form of 'Prithvi' (Earth), Agni (Fir), Jal (Water), Vayu (Air) and Akash (Space) were all worshipped or revered in one of other form from the ancient times as per Hindu mythology. These elements of nature were visible in the form of air, water, soil, flora and fauna that came to be regarded as the abodes of gods or goddesses and as a result got protection for spiritual, religious cultural and social reasons (*cf.* Patil, 2020). An ancient quote says: "It was through the worship of trees that man attempted to approach God". Multiple utility of plants was realized from early times and this made these as the one of the oldest form of veneration. The sacred trees are worshiped. There is some specific festival or sentiments associated with the specific reason.

## Taboo for Plant Conservation

From the current study it was found that in Jalgaon district people believed that planting a five trees in life gives Moksha (to live in heaven after the death). Tree plantation is common by VIP's during their visit to various establishments. Offering of saplings to guest of some functions is also a custom in this region. Planting a sacred tree in memory of relatives is also novel concept in this region. Such planted area is called as 'Smrutivan'. Taboos are the unwritten, orally transmitted traditional and social rules that regulate human behavior (Colding and Folke, 1997). In rural areas many plants and animals and considered as pious and sacred. The religious belief serves as an instrument of protection of those rare species (Pandey *et al.*, 2003).

## Beliefs and Myth Towards Soil and Water Conservation

Sacred plants present in vicinity of temples improve the soil stability of the region and act as soil binder. Grasses and *Eucalyptus* species are maintained to bind soil thereby preventing soil erosion. Water of different rivers in this region such as Tapi, Girna, Bori, Panzara are treated as holy. People bath at the confluence of two or three rivers called 'Sangam' (union) in different part of the region, *e.g.* Tapi sangam at Changadev (Jalgaon district) and Prakasha (Nandurbar district). Small natural water bodies are maintaining near sacred trees. Hot springs are also present near sacred plants, *e.g.* Unapdev hot sprint (Jalgaon district).

## Conservation of Medicinal and Aromatic Plants through Local Beliefs

Near the vicinity of temple holy places or near the sacred forest important medicinal and aromatic plants are present. Most of the tribal such as Pawara, Bhil, Tadavi living in the forest are illiterate. They have nurtured traditional customs, rituals, ceremonies and way of forest life through folk belief with great vigor Poojari, Bhagat and Buwa are advise to treat people. This traditional knowledge inherited orally from one generation to other. *Aloe vera*

L., *Chlorophytum borivilianum* Sant. and Frenand are now planted by the tribals in Chopda and Raver area. Development of 'Landorkhore' project of medicinal plants at Jalgaon, medicinal plant project at Toranmal, Mangal Dev project at Amalner are attempts at conservation of medicinal plants.

## Conservation of Floral Diversity

Many plants or trees species are accepted as sacred by Hindus and hence worshipped. Due to deep rooted religious system, many trees are not axed as they are supposed to increase longetivity. In Khanesh, *Ficus religiosa* L., *Ficus bengalenisis* L., *Ficus racemosa* L., *Ocimum tenuiflorum* L., *Calotropis gigantea* (L.) R. Br. have religious significance. The Neem tree is considered to be a manifestatum of goddess Durga. Tulsi plant (*Ocimum tenuiflorum* L.) is worshipped as goddess by women throughout the country which is supposed to enhance the longevity of their husband's life.

In a nutshell, the present investigators initiated to document traditional uses of plants species found within the courtyards of temples. Some temples have no definite boundary and have better surrounding land which is generally covered by wild plants species. This is particularly so in case of temples or deities in tribal region in this part of Maharashtra. These sacred plants and those found within the ambit of temples and adjacent lands are tapped for their concrete as well as abstract uses in the region. Plants when associated intimately with human societies as in the present cases of temple yards are subjected more rigorously by the ambient communities. It was, therefore, possible to gather information on medicinal and miscellaneous purposes, apart from sacred and socio-economically useful plant species.

During botanistion and particularly while carrying out floristic or biodiversity studies, temple yards and adjacent lands are overlooked by the investigators, although these plant groups are easily accessible. The present authors made a headway in this realm of research (Khare *et al.*, 2021a,b; 2022). Moreover, plants are studied with principal emphasis for their concrete or material uses and abstract relationships of mankind with the plant world are ignored. The present investigator, therefore, laid more particular emphasis on such neglected plant species to pinpoint their (hidden) potentialities for the welfare of local communities and even entire human societies. Such an investigation may filip local economy and guide for underutilized plant species.

Indigenous cultures are rich repositories of valuable herbal practices and plant-based utilities that help sustain life of local people. In present time, the rate of acculturation in rural and tribal areas is rather faster as compared to the past because of rapid forces like urbanization, liberalization, modernization and globalization. These forces not only deprive mankind from the potentialities of the ambient biodiversity but also lead to cultural erosion. It is the cultural erosion that accelerates knowledge erosion of indigenous societies. Cultural and spiritual values are also changing and thus the existing bond between the people and nature is weakened. It was, therefore, thought imperative to tap traditional knowledge and potentialities in areas wherein the research workers have largely neglected. It is hoped that the present account may fill in the observed lacunas in the region under study. Many useful plant species which are an integral part of the multi-species cover have become endangered. Some of these find places in plantation purposely in sacred places. We should also pay attention on scientific ground.

An inventory on the scared and socio-religious plant species as employed by people of Khandesh region was made. It is the outcome of botanization for about four years in the said region. The investigators visited the temples of various deities worshipped by urban, rural and tribal folks in the region. They also made surveys of social envents like rites, rituals, ceremonies, festivals, pilgrims *etc.* for the plants taxa employed by the Indigenous people. Information was tapped during personal visits to such places and religious occasions to obtain first-hand information. The plant specimens or parts were collected to decipher their specific identity by using national, state, regional and district floras. Some places of interest are photographed. Discussions with concerned informants were held during spot visit. Data regarding plant or part used, local plant name, purpose for which plants employed was critically and carefully documented in field diary. Repeated visits helped to verify the information adduced during earlier visits, faith, belief, tradition, customs *etc.* were acknowledged while interacting with the indigenous people. The results of in-depth study in different seasons and communities during the tenure of field study are presented in this communication. The investigators completed research on this much overlooked area of research especially in Dhule, Nandurbar and Jalgaon districts of Khandesh.

This communication, as precursory to this nature of topic of investigation, provides background information with respect to geographical position, rainfall, temperature, climate, forests, and crops, people *etc.*, apart from earlier literature. The reasons for undertaking this topic of research are explained. A standard methodology was adopted by investigators to tap down the information from the knowledgeable persons of all castes, tribes and religion observed in these districts.

Floristic analysis of the taxa investigated revealed as many as 263 plant species belonging to 221 genera and 83 families, a single species being a gymnosperm as enumerated in Table 1. Of these, the dicotyledons are represented by 232 species belonging to 194 genera and 73 families, whereas the 31 monocotyledonous species belong to 27 genera and 10 families. This plant taxa are further categoried into habital groups *viz.*, trees (89), shrubs (49), climbers (30), and herbs (96).The figures in parenthesis stand for the number of species recorded in the region and places under study. Trees especially belonging to monocotyledons are only 04 species. The overall data indicated that the dicotyledons play a major role in indigenous and traditional utilities. Totals

103 species are found under cultivation, while rest others run wild in region or places visited. Some of the wild species are also observed under cultivation. This fact indicated that the people of Khandesh have endeavored to domesticate local plant species for various purposes and brought them ultimately under traditional cultivation. It is also a noteworthy fact that about 57 species are exotic and belonged originally to different parts of the world. Interestingly, maximum exotic species (27 species) belonged to American continent. Their probable period of introduction or invasion in the Indian subcontinent is critically discussed based on the evidences such as ancient Sanskrit literature through epics, Puranas and so on, apart from anthropology, archeology, ethnobotany, history, ancient trade, palaeobotany *etc.* Their role is assessed in the local economy. Some of the exotic species are discussed at proper length.

The traditional utilities have been broadly classified as: (i) sacred uses (ii) medicinal uses and (iii) miscellaneous purposes. In each category, they are evaluated floristically besides the plant parts and status of species, whether wild or cultivated. Total 232 species are found medicinally useful. Total 143 species are employed for sacred and religious purposes by the urban elites, rural and tribal folks in the region. Another category called miscellaneous purposes include total 44 plant species, whether wild or cultigens. In every category of utilities, the dicotyledons constituted a major share in the sustenance of the local people.

The role of local people is discussed for the sake of conservation of local biodiversity in various forms of traditional employment. Some sacred *vis-à-vis* socio-religious plant taxa interpreted and compared with the ancient literature like Rigveda. The plants species used for various purposes are gathered whenever available in some seasons throughout a year. However, some plants parts or entire parts are sold at certain places of worships, temples, pilgrims, on occasions of festivals or ceremonies. They are particularly highlighted. After total observations and experiences in the region, the investigators tried to seek a special attention to have gardens by the trust of the temples or individuals of the needy societies. Some common species of great demand are earmarked and suggested in the form of a model garden. This suggested activity may help uplift the economy and business of the local people. Lastly, the results, observations and experiences based on the first-hand information are discussed relevantly suitable to the needs of local populace. The overall planned research with clear objectives is supported by necessary tables, appendices, photographs following standard methodology. All relevant and recent references have been arranged alphabetically and chronologically at the end of this communication.

## An Appeal

India is a land with unique amalgam of castes, races, religions, cultures, languages, dialects and also enriched by biological and ethnobiological diversities. It is thus a supreme symbol of culture. The fire particularly of ethnobotanical researches has attracted many Indian researchers. Although so, biodiversity and ethnobotanical wisdom associated with the temple courtyards or holy places have not been paid due attention they deserve. The present authors, therefore appeal to have an eye on this inviting area of research.

## Appendix I
## Temples and Deities in Khandesh

| Sl.No. | Deity | Place | Tehsil |
|---|---|---|---|
| **(I) Jalgaon District** | | | |
| 1. | Bhavani Mandir | Wade | Bhadgaon |
| 2. | Chakradhar Swami | Kanashi | Bhadagaon |
| 3. | Changdev | Chandev | Muktainagar |
| 4. | Dandapaneshwar | Nandurbar | Nandurbar |
| 5. | Datta Mandir | Chorwad | Parola |
| 6. | Devmogaradevi Mandir | Visarwadi | Navapur |
| 7. | Dwarja Devi | Bahal | Chalisgaon |
| 8. | Hanuman Mandir | Shirsale | Bodawad |
| 9. | Hartale | Hartale | Muktainagar |
| 10. | Indagiri | Nagaon | Amalner |
| 11. | Kapileshwar | Amalner | Amalner |
| 12. | Kartiki Swami | Anturli | Amalner |
| 13. | Mahadev Mandir | Valziri | Chalisgaon |
| 14. | Mahadev Mandir | Vaghra | Parola |
| 15. | Maheji Devi | Maheji | Erandol |
| 16. | Mahiji Devi | Patonda | Amalner |
| 17. | Mangalgraha | Amalner | Amalner |
| 18. | Manudevi | Manudevi | Yawal |
| 19. | Mary-Aai | Wade | Bhadgaon |
| 20. | Mudhai devi | Vaghali | Chalisgaon |
| 21. | Muktai | Kothali | Muktainagar |
| 22. | Munjoba | Atrawal | Yawal |
| 23. | Padmalay | Mhasawad | Erandol |
| 24. | Patanadevi | Patana | Chalisgaon |
| 25. | Ram mandir | Wade | Bhadgaon |
| 26. | Saptashrungi Mata | Shiragad | Chopada |
| 27. | Sati Mata | Chandani Kurhe | Amalner |
| 28. | Shivdham | Ratnapimpri | Parola |
| 29. | Unapdev | Unapdev | Chopadaa |
| 30. | Vaghreshwari mata | Nandurbar | Nandurbar |
| **(II) Dhule District** | | | |
| 1. | Ashapuri Devi temple | Patan | Dhule |
| 2. | Bhatai Devi | Ner Kusumba | Sakri |
| 3. | Datta Mandir | Sarangakheda | Shindkheda |

| Sl.No. | Deity | Place | Tehsil |
|---|---|---|---|
| 4. | Dhanadai Mata | Mhasadi | Sakri |
| 5. | Dhandai Mata Temple | Navalnagar | Dhule |
| 6. | Ekvira Devi | Dhule | Dhule |
| 7. | Jogeshwari Mata | Jogshelu | Shindkheda |
| 8. | Jogeshwari Mata Mandir | Shindkheda | Shindkheda |
| 9. | Mhalsa | Ardhe | Shirpur |
| 10. | Nagai Mata | Kokale | Sakri |
| 11. | Pedkai Mata | Vikharan | Shindkheda |
| 12. | Saptashrungi Mata and Jogeshwari Mata | Betawad | Shindkheda |
| 13. | Sati Devi | Boris | Dhule |
| | **(III) Nandurbar District** | | |
| 1. | Ashtabhuja Devi | Mandane | Shahada |
| 2. | Astambha Rhishi | Astambha | Akkalkuwa |
| 3. | Bhatai Devi | Ner Kusumba | Sakri |
| 4. | Dandapaneshwar | Nandurbar | Nandurbar |
| 5. | Datta Mandir | Sarangakheda | Shindkheda |
| 6. | Devi Mogra Mata | Malda-Mogra | Taloda |
| 7. | Devmogaradevi Mandir | Visarwadi | Navapur |
| 8. | Dhandai Mata | Mhasadi | Sakri |
| 9. | Gorakhnath Temple | Toranmal | Dhadagaon |
| 10. | Gulam Maharaj | Ranjanpur | Taloda |
| 11. | Jogeshwari Mata Mandir | Shindkheda | Shindkheda |
| 12. | Mahanubhav Panth Temple | Hingani | Shahada |
| 13. | Manudevi | Manudevi | Yawal |
| 14. | Nagai | Kokale | Sakri |
| 15. | Prakasha | Prakasha | Shahada |
| 16. | Rokadmal | Rokadmal | Shahada |
| 17. | Saptashrungi Mata | Betawad | Shindkheda |
| 18. | Shree Ganesha (Heramb) temple | Jaynagar | Shahada |
| 19. | Umaj Mata temple | Shinda | Nandurbar |
| 20. | Vaghreshwari mata | Nandurbar | Nandurbar |

## Appendix II
## Festivals In Khandesh

| Sl.No. | Festival name | Gregorian calendar | Deity |
|---|---|---|---|
| 1. | Akshay Trutia | April | Varied |
| 2. | Ashadi Ekadashi | Jun-July | Vitthal |
| 3. | Bakari Eid | Varied | Esmail-ule-Salam |
| 4. | Bali-Pratipada | October–November | Bali |
| 5. | Bhaldev | September- October | Gorakshanath |
| 6. | Bhaubij | October–November | Varied |
| 7. | Bhogi | January | Varied |
| 8. | Buddha Pournima | May | Buddha |
| 9. | Christmas | December | Jesus |
| 10. | Dasehara | September October | Varied |
| 11. | Easter | April | Yeshu |
| 12. | Eid-Ul-Fitar | Varied | Moh. Paigambar |
| 13. | Ganesh utsav | August–September | Ganapati |
| 14. | Ghat Sthapana | September -October | Durga Mata |
| 15. | Gudi Padwa | March–April | New Year |
| 16. | Gulabai | September- October | Mahadev-Parvati |
| 17. | Guru Nanak Jayanti | October- November | Guru Nanak Jayanti |
| 18. | Guru Pournima | July | Datta Guru |
| 19. | Hanuman Jayanti | April | Hanuman |
| 20. | Haratalika | August- September | Mahadev-Parvati |
| 21. | Holi | March | Holika |
| 22. | Janmashtami (Shri Krishna jayanti) | August -September | Krishna |
| 23. | Kojagiri Pournima | September–October | Laxmi |
| 24. | Laxmi Pujan | October–November | Laxmi |
| 25. | Maha Shivaratri | January–February | Shiva |
| 26. | Mahalaxmi | August- September | Mahalaxmi |
| 27. | Mahalaxmi | July-August | Mahalaxmi |
| 28. | Mahavir Jayanti | April | Mahavir |
| 29. | Mahavir Jayanti | April | Bhagwan Mahavir |
| 30. | Makar Sankranti | January 14 or 15 | Varied |
| 31. | Mangala Gaur | August | Shiva |
| 32. | Moharam | Varied | Imam Husain |
| 33. | Nagpanchami | July-August | Snake God |
| 34. | Narak Chaturdashi | October–November | Varied |
| 35. | Navaratri | September–October | Durga |

| Sl.No. | Festival name | Gregorian calendar | Deity |
|---|---|---|---|
| 36. | Pateti | August | Varied |
| 37. | Pola | August–September | Bullock Worship |
| 38. | Rakshabandhan | July -August | Varied |
| 39. | Ram Navami | March–April | Rama |
| 40. | Rangapanchami | March | Varied |
| 41. | Rhishi Panchami | | Guru Datta |
| 42. | Shivaji Maharaj Jayanti | February | Shivaji Maharaj |
| 43. | Tij | January | Varied |
| 44. | Vasant Panchami | January-February | Varied |
| 45. | Vasu Baras | October-November | Varied |
| 46. | Vatasavitri Pournima | August–September | Worship of Baniyan Tree |

## Appendix III: Tribal Festivals in Khandesh

| *Sl.No.* | *Festival name* | *Gregorian calendar* | *Deity* |
|---|---|---|---|
| 1. | Astambha | October-November | Ashwasthama |
| 2. | Bhongarya Bajar | March | Varid |
| 3. | Dasehara | September October | Varied |
| 4. | Dindan | | |
| 5. | Diwali | October–November | Varied |
| 6. | Dongarya Dev | | Dongarya Dev |
| 7. | Gav-Diwali | October-November | Varid |
| 8. | Goddess Kansari Festival | At the time of Harvesting | Goddess Kansari |
| 9. | Gulalya Bajar | | Varid |
| 10. | Hivarya Dev | | Hivarya Dev |
| 11. | Holi | March | Holika |
| 12. | Indal | | Indra Dev |
| 13. | Indraja | Lucky day between Dasra and Divali | Indra |
| 14. | Khamb Dev | | Khamb Dev |
| 15. | Maha Shivaratri | January–February | Shiva |
| 16. | Nagdivali | January | Snake god |
| 17. | Nagpanchami | July-August | Snake God |
| 18. | Nai-Chari | | |
| 19. | Nandario dev festival | beginning of rainy season | Nandario god |
| 20. | Navay | | |
| 21. | Paludya Dev (Vilichari) | | |
| 22. | Rakshabandhan | July -August | Varied |
| 23. | Rangapanchami | March | Varid |
| 24. | Waghdev | August | Tiger god |

## Appendix IV
## Pilgrims In Khandesh

| Sl.No. | Deity | Place | Period | Tehsil |
|---|---|---|---|---|
| 1. | Astambha Rhushi | Astambha | During Diwali Period | Akkalkuwa |
| 2. | Bhatai Devi | Ner Kusumba | | Sakri |
| 3. | Chakradhar Swami | Kanashi | Guru Pournima | Bhadagaon |
| 4. | Chakradhar Swami | Kanashi | Datta Jayanti/Rhishi Panchami | Bhadgaon |
| 5. | Changdev | Chandev | January-February | Muktainagar |
| 6. | Datta Guru | Chorwad | December | Parola |
| 7. | Datta Mandir | Sarangakheda | Datta Jayanti | Shindkheda |
| 8. | Devmogaradevi Mandir | Visarwadi | Mahashivratri | Navapur |
| 9. | Ekvira Devi | Dhule | Narali Pournima | Dhule |
| 10. | Gorakhnath | Toranmal | Nandurbar | Dhadgaon |
| 11. | Hanuman Mandir | Shirsale | Hanuman Jayanti | Bodawad |
| 12. | Jogeshwari Mata Mandir | Shindkheda | Navaratri | Shindkheda |
| 13. | Kapileshwar | Amalner | Mahashivratri/ Shravan Month Monday | Amalner |
| 14. | Kartiki Swami | Anturli | Kartiki Pournima | Amalner |
| 15. | Khandoba | Nagardeola | January-February | Pachora |
| 16. | Khopadi | Shahapur | November | Amalner |
| 17. | Mahadev | Nandagaon | Mahashivratri | Bodawad |
| 18. | Mahadev Mandir | Valziri | Mahashivratri | Chalisgaon |
| 19. | Mahadev Mandir | Vaghra | Mahashivratri | Parola |
| 20. | Manudevi | Manudevi | Navaratri | Yawal |
| 21. | Mudhai devi | Vaghali | Navaratri | Chalisgaon |
| 22. | Muktai | Kothali | Mahashivratri | Muktainagar |
| 23. | Muktai | Mehun | Mahashivratri | Muktainagar |
| 24. | Munjoba | Atrawal | January-February | Yawal |
| 25. | Padmalay | Mhasawad | Ganesh Utsav/ Mangali Chaturthi | Erandol |
| 26. | Patanadevi | Patana | Navaratri | Chalisgaon |
| 27. | Prakasha | Prakasha | | Shahada |
| 28. | Rhishi Pantha | Rashid Pantha | Rhishi Panchami | Chalisgaon |
| 29. | Sakharam Maharaj | Amalner | May | Amalner |
| 30. | Saptashrungi Mata | Shiragad | Navaratri | Chopada |

| Sl.No. | Deity | Place | Period | Tehsil |
|---|---|---|---|---|
| 31. | Saptashrungi Mata | Betawad | Navaratri/Chaitra Pournima | Shindkheda |
| 32. | Satimata | Chandani Kurhe | January | Amalner |
| 33. | Shivdham | Ratnapimpri | Mahashivratri | Parola |
| 34. | Unapdev | Unapdev | Mahashivratri | Chopadaa |

## Appendix V
## Balaji Rath Yatra

| Sl.No. | Deity (Rath) | Place | Tehsil |
|---|---|---|---|
| 1. | Balaji | Betawad | Shindkheda |
| 2. | Balaji | Bahal | Chalisgaon |
| 3. | Balaji | Bahadarpur | Parola |
| 4. | Balaji | Parola | Parola |
| 5. | Balaji | Khedgaon | Chalisgaon |
| 6. | Ram | Jalgaon | Jalgaon |
| 7. | Balaji | Ratnapimpri | Amalner |
| 8. | Balaji | Chorwad | Parola |
| 9. | Balaji | Chalisgaon | Chalisgaon |
| 10. | Balaji | Chopada | Chopada |
| 11. | Balaji | Shendurni | Jamner |
| 12. | Sakharam Maharaj | Amalner | Amalner |
| 13. | Navaratri Vhan | Dharangaon | Dharangaon |

## Appendix VI
## Gramdevta (Village Deities) commonly found near villages

| Sl.No. | Gramdevta | Sl.No. | Gramdevta |
|---|---|---|---|
| **1.** | Khandoba | 2. | Khandoba |
| 3. | Khandoba | 4. | Mhasoba |
| 5. | Mhalasa | 6. | Matoba |
| 7. | Pir | 8. | Munjoba |
| 9. | Viroba | 10. | Bhairoba |
| 11. | Asara | | |

## Appendix VII
## Some Symbolism And Culture Observed In Khandesh

| *Sl.No.* | *Symbolism and Culture* | *Sl.No.* | *Symbolism and Culture* |
|---|---|---|---|
| 1. | ओंकार (ॐ) | 14. | Kamal |
| 2. | Swastik | 15. | Vriksha Pujan |
| 3. | Kar-Darshan | 16. | Audumbar |
| 4. | Pratah Smaran | 17. | Tulas |
| 5. | Angan Puja | 18 | Wat-Vriksha |
| 6. | Kalash | 19 | Amra-Vriksha |
| 7. | Rangoli | 20 | Kel |
| 8. | Kunku-Tilak | 21 | Awali-bhijan |
| 9. | Gandha | 22 | Amrut Vriksha |
| 10. | Deep | 23 | Pimpal |
| 11. | Arati | 24 | Trishul |
| 12. | Shriphala | 25 | Supari |
| 13. | Sanjvat | | |

## Appendix VIII
## Some Rites (Vrat) Observed In Khandesh

| *Sl.No.* | *Vrat* |
|---|---|
| 1. | Mas Vrat (From months Chaitrya to Falgun) |
| 2. | Danvrat |
| 3. | Grahaswami (Based on date of birth) |
| 4. | Vara-nusar (Based on the days of weak) |
| 5. | Rashi-Vrat (Based on sign of zodiac) |
| 6. | Ekadashi Vrat |
| 7. | Vrat during Chaturmas |
| 8. | Vrat of Adhik mas |
| 9. | Vrat observed on Thursday in Margashirsh month |
| 10. | Shri. Mahalaxmi Vrat |
| 11. | Sadesati Vrat |
| 12. | Sola (Sixteen) Somvar (Monday) Vrat. |

# References


Acosta Christabal (1578) tractado de las drogusy medicinas de las Indias Orientalis, Burgos.

Bailey L.H. (1949) Manual of Cultivated Plants (Rev.Ed.) The Macmillan Co., New York, USA.

Balick, M.J. (1996) Transforming ethnobotany for the new millennium. Ann. Missouri Bot. Gard. 66: 58-66.

Bodding, P.O. (1925) Studies in the Santal medicine and connected folklore- I. The Santals and disease. Mem. Asiat. Soc. Bengal 10 1(1) : 1-132.

Bodding, P.O. (1927) Studies in the Santal medicine and connected folklore- II. Santal medicine. Mem. Asiat. Soc. Bengal 10 (2) : 133-426.

Bodding, P.O. (1940) Studies in the Santal medicine and connected folklore. III. Mem. Asiat. Soc. Bengal 10 (3) : 427-502.

Colding, J. and C.Folke (1997) The relations among threatened species, their protection and taboos. Conservation Ecology 1(1):6.

Dadhich Shubha and N.K.Dwivedi (2017) Conservation of plant biodiversity in temple yards of Luni Subdistrict, Jodhpur, Rajasthan, India. IOSR Journal of Pharmacy And Biological Sciences 12(5):31-39.

Daniel, F.Austin (2008) Evolvulus alsinoides (Convolvulaceae): An American herb in the Old World. Journal of Ethnopharmacology 117(2):185-198.

Debnath A. and Debnath B. (2017) Diversity, invasion status and usages of alien plant species in Northeastern Hilly State of Tripura: A confluence of Indo-Barman Hotspot. American Journal of Plant Sciences 8:212-235.

Dogra, K.S. (2011) Alien plants distribution and ecology in the temple courtyards of Himachal Pradesh (N.W.Himalaya). Himachal Prdesh University Journal, July:1-11.

Faulks, P.J. (1958) An Introduction To Ethnobotany. BCM/Moredale. Moredale Publications Ltd. London W.C. 1.

Geeta, R. And Gharaibeh Waleed (2007) Historical evidences for pre-Columbian presence of Datura in the Old World and implications for a first millennium transfer from the New World. Journal of Biosciences 1: 1227-1244.

Ghate, V.S. (1998) Plants in 'Patra-Pooja': notes on their identity and utilization. Ethnobotany 10:6-15.

Gupta Shakti M. (1996) Plants In Indian Temple Art. B.R. Publishing Corporation, Delhi, India.

Harshberger, J.W. (1895) Some new ideas. Philadelphia Evening Telegraph.

Heiser, C.B. (1995) Ethnobotany of Niranjilla. Solanum quitosense and its relatives. Bot. Gaz. 36: 4-11.

Hooker, J.D. (1872-1897) The Flora of British India. Vol.I-VII. L.Reeve and Co., London, UK.

Hooker, J.D. and T.Thompson (1883) An Introductory Essay To The Flora Of India. W. Pamplin, 45, Frith street, Soho, London, UK. (Repr.Ed.1855).

Jain, S.K. (2001) Ethnobotany in modern India. Phytomorphology 51 (Golden Jubilee): 39-54.

Jain, S.K. (2015) Dictionary of Indian Folk Medicine And Ethnobotany. Deep Publications, New Delhi, India.

Johannessen, Carl L. and Anne Z. Parkar (1989) Maize ears sculpted in 12th and 13th century A.D.India as indicators of Pre-Columbian diffusion. Economic Botany 43 (2): 164-180.

Johannessen, Carl L. and Wang Siming (1998) American Crop Plants in Asia before AD 1500. Pre-Columbian: A Journal of Long-Distance Contacts 1:9-36.

Jones, V.H. (1941) The nature and scope of ethnobotany. Chronica Botanica 6 (10) : 215-221.

Khare, S., Pawar, S., Vishwakarma, K.S. and D.A.Patil (2014). Ethnobotanical notes of certain medicinal plants used by tribals of Nandurbar District (Maharashtra), India. International Journal of Biodiversity and Biosciences 4(3):200-202.

Khare, S.M., Pawar Shubhangi and D.A.Patil (2021a) Ethnobotany of indigenous plants in Khandesh region (Maharashtra: India). International Journal of Botany Studies 6(5):693-698.

Khare, S.M., Pawar Shubhangi and D.A.Patil (2021b) Plants of temples and religious places in Khandesh region (Maharashtra): An Ethnobotanical Perspective. Plantae Scientia, 4(4 and 5):257-265.

Khare, S.M., Pawar Shubhangi and D.A. Patil (2022) Observations on Bioculture and Medicinal Practices in Akkalkuwa Tehsil (Nandurbar District) In Maharashtra. Plantae Scientia 05(01):4-7.

Kshirsagar, S.R. and D.A.Patil (2008) Flora of Jalgaon District (Maharashtra). Bishen Singh, Mahendra Pal Singh, Deharadun, India.

Mahabale, T.S. and C.R. Karnik (1960) Distribution of structure of frests in Satpura ranges in Khandesh districts and factors governing them. J.Univ. Bombay 28 (3):59-77.

Manilal, K.S. (2007) The Hortus Malabaricus : An account of heritage plants of Malabar. Ethnobotany 19 : 17-31.

Martin, F.W. Campbell, C.W. and Ruberte, R.M. (1987) Perennial Edible Fruits of The Tropics: An Inventory. U.S. Department of Agriculture. Agriculture Handbook No.642, 222 p.illus.

Mashelkar, R.A. (2002) In Forward. Indian J. Trad. Knowledge. 1 (1) : 1

Medakkar, S.S. and P.P. Sharma (2016) Antiquity of some exotic plants in India. The South Asian Academic Research Chronicle III (6):6-14.

Mohanty Rajballav, Tripathy, B.K. and T.Panda (2012) Role of temples and other holy places in plant conservation of Odisha, India. International Journal of Conservation Science 3(4):301-308.

Mohanty, R.B., Mohapatra, B.K. and S.N. Padhy (1997) Plant conservation in temple yards of Orissa. Ancient Science of Life 17(2):94-99.

Mukhopadhyay, D.P. and R.K. Chakraverty (2008) Plant Wealth of The Raj Bhavan Kolkata. Occassional Paper-5 from Raj Bhavan, Kolkata, March, 2008.

Naik, V.N. (1998) Flora of Marathwada. Vol.I-II. Amrut Prakashan, Aurangabad (Maharashtra), India.

Nandita Krishna and M. Amirthalingam (2014) Sacred Plants of India. Penguin Random House, Gurgaon, India.

Pandey, H.N., Tripathi, O.P. and Tripathi, R.S. 2003. Ecological analysis of forest vegetation of M eghalaya. In: Approaches for Increasing Agriculture Productivity in Hill and Mountain Ecosystem (Ed. Bhatt, B.P., Bujarbaruah, K.M., Sharma, Y.P. and Patiram). ICAR Research Complex for N EH Region, Umain, Meghalaya. Pp.37-49.

Patil, D.A. (1992) Key To The Angiospermic Fanilies of North Maharastra. Pune Vidyarthi Prakashan Griha, Pune (M.S.) India.

Patil, D.A. (2003) Flora of Dhule and Nandurbar Districts (Maharashtra). Bishen Singh, Mahendra Pal Singh, Deharadun, India.

Patil, D.A. (2011) Ethnomedicine to modern medicine : Genesis through ages. J. Exp. Sci. 2 (3), 25-29.

Patil, D.A. (2017) Invasive alien species in Khandesh region (Maharashtra, India): Diversity, implications and measures. *Sch.Acad.J.Biosci.* 5(12):867-876.

Patil, D.A. (2019) Food Crops: Evolution, Diversity And Advances, Scientific Publishers, Jodhpur, India.

Patil, D.A. (2020) Indian Plantlore: Beliefs, Myths and Legends. Lambert Academic Publishing, Beau. Bassin, Mauritius.

Patil, D.A. (2021) Plant Invasion In India. Lambert Academic Publications, Beau Bassin, Mauritius.

Patil, D.A., Patil P.S., Dushing Y.A., Aher U.P. and Y.A. Ahirrao (2010) Ethnobotany of Buldhana District (Maharashtra), India. Daya Publishing House, Delhi, India.

Patil, M. V., Shubhangi Pawar, and D. A. Patil. (2006) Ethnobotany of Butea monosperma (Lam.) Kuntze in North Maharashtra, India. Journ.Non-Timber For.Product 13(2):131-133.

Patil, M.V. and D.A.Patil (2006) Ethnobotany of Nasik District (Maharashtra). Daya Publishing House, Delhi, India.

Patil, S. L., and D. A. Patil (2007) Ethnomedicinal plants of Dhule district, Maharashtra. Natural Product Radiance (CSIR) 6(4):341-346.

Pawar Shubhangi and D.A. Patil (2010) Identities and utility of botanicals used in religious worships in Khandesh region of Maharashtra. Ethnobotany 22:102-106.

Pawar Shubhangi and D.A.Patil (2008) Ethnobotany of Jalgaon District (Maharashtra). Daya Publications House, Delhi, India.

Prabhakran, R. and G.Sabari Lakshmi (2017) Studies on sthalavrikshas of various temples in Tamil Nadu, India. Bioscience Discovery 8(1):64-72.

Pullaiah, T. and D.Murlidhara Rao (2002) Flora of Eastern Ghats. Vol.1. Regency Publications, New Delhi, India.

Pullaiah, T. and K. Sri Ramamurthy (2001) Flora of Eastern Ghats. Vol.2. Regency Publications, New Delhi, India.

Puri, G.S., Jain, S.K. and U.R.Deshpande (1960) Observations on grasslands of Khandesh. Gosamvardhana 8(6):19-22.

Rheede, H.A. Van (1678-1693) Horti India Malabarici. 12 Volumes. Amsterdam.

Robbins, W.W.; Harrington, J.P. and B.Freire-Marreco (1916) Ethnobotany of the Tewa Indians. Bureau of American of Ethnology. Smith-Soman Institution. Bull. 55:1-124.

Roxburgh, W. (1832) Flora India. Today and Tomarrow, New Delhi, India. (1971 Repr.Ed.).

Saraswat, K.S. (1992) Archaeobotanical remains in ancient culture and socio-economical dynamics of the Indian subcontinent. Palaeobotanist 40:514-545.

Saraswat, K.S. (2005) Agriculture background of the early farming communities in the Middle Gangetic Plain. Praghdhara 15:145-177.

Shah, N.C. (2008) Prospects of ethnobotany in India. Ethnobotany 20 : 16-24.

Sharma, B.D., Karthikeyan, S. and N.P.Singh (1996) Flora of Maharashtra State: Monocotyledones. Bot.Surv. India. Calcutta, India.

Sharma, U.K. (2019) Ethnomedico-religious Plants of Assam. Bishen Singh Mahendra Pal Singh, Dehradun, India.

Sheikh, D.K. and A.K. Dixit (2017) Occurrence of invasive plants in three phytogeographical region of Bilaspur district of Chhatisgarh. Annals of Plant Sciences 6(12):1872-1878.

Singh Arjun (2014) Temple architecture and iconography of Babour and Panjnara temples in Jammu region. Munich, GRIN Verlag.

https://www.grin.com/document/284488.

Singh Th. B., Das, A.K. and P.K.Singh (2015) Study of alien and invasive flora of Valley District of Manipur and their control. International Journal For Innovative Research In Science and Technology 1(12):2349-601.

Singh, A.K. and S.N.Nigam (2017) Ancient alien crop introductions integral to Indian agriculture: An overview. Proceedings of Indian National Science Academy 83 (3): 549-568.

Singh, K.A. and Suchita Sharma (Ed.) (2008) Temple Architecture of The Western Himalayas: Ravi And Beas Valley. Sundeep Prakashan, Delhi, India

Singh, N.P., Lakshminarsimhan, P., Karthikeyan, S. and P.V.Prasanna (2000) Flora of Maharashtra State: Dicotyledons. Vol.I. Bot.Surv.India, Calcutta, India.

Singh, N.P., Lakshminarsimhan, P., Karthikeyan, S. and P.V.Prasanna (2001) Flora of Maharashtra State: Dicotyledons. Vol.II. Bot.Surv.India, Calcutta, India.

Singh, S.C. and G.N. Srivastava (2000) Exotic medicinal plants of Lucknow district (U.P.) India. In: Ethnobotany And Medicinal Plants of Indian Subcontinent (Ed.J.K.Maheshwari) Scientific Publishers, Jodhpur, India. Pp.223-235.

Sood, S.K., Thakur Vandana and T.N. Lakhanpal (2005) Sacred And Magico-religious Plants of India. Scientific Publishers, Jodhpur, India.

Stewart, R.R. (1972) An Annonatal Catalogue of The Vascular Plants of West Pakistan And Kashmir. Fakhri Press, Karachi, Pakistan.

Struwig, M. and S.W. Siebert (2013) A taxonomic revision of Boerhavia (Nyctaginaceae) in Southern Africa. South African Journal of Botany 86:116-134.

Vartak, V.D. and Madhav Gadgil (1981) Studies on sacred groves along the western ghats from Maharashtra and Goa: Role of beliefs and folklores. In: Glimpses of Indian Ethnobotany (Ed.S.K.Jain) Oxford and IBH Publishing Co; New Delhi, India. pp.272-278.

Vartak, V.D. and Madhav Gadgol (1990) Studies on sacred groves along the western ghats from Maharashtra and Goa: Role of beliefs and folklores. In: Contributions To Ethnobotany of India (Ed.S.K.Jain) Scientific Publishers, Jodhpur, India. pp.329-335.

Vishnu-Mittre, (1974) Palaeobotanical evidences in India. Evolutionary Studies In World Crops (Ed.J.B.Hutchinson) Cambridge Univ. Press Cambridge UK.

Watt, G. (1889-1896) A Dictionary of Economic Products of India. Vol. 1-6. Periodical Expert, Shahadara, Delhi, India.

Watt, G. (1908) Commercial Products of India : Being an Abridgment of Dictionary of Economic Products of India, London, UK.

# Index

9 789359 199948

Printed at Repro India Ltd.